Johannes Hevelius, Charles Leeson Prince

The Illustrated Account

Given by Hevelius in His Machina Celestis of the Method of Mounting His

Telescopes and Erecting an Obsavertory

Johannes Hevelius, Charles Leeson Prince

The Illustrated Account
Given by Hevelius in His Machina Celestis of the Method of Mounting His Telescopes and Erecting an Obsavertory

ISBN/EAN: 9783337251376

Printed in Europe, USA, Canada, Australia, Japan

Cover: Foto ©berggeist007 / pixelio.de

More available books at **www.hansebooks.com**

Ætheris et vasti Mensor Celebratus Olympi,
Hunc vultum HEVELIUS MAGNUS et ora gerit,
Conscia cælesti dispensans Sidera mente,
Grande micat Patriæ Sidus et Ipse Suæ.

THE

ILLUSTRATED ACCOUNT

GIVEN BY

HEVELIUS

IN HIS

"MACHINA CELESTIS"

OF THE METHOD OF MOUNTING

HIS TELESCOPES

AND ERECTING AN

OBSERVATORY.

Reprinted from an Original Copy,
with some Remarks,

BY

C. Leeson Prince,

Member of the Royal College of Surgeons,
Fellow of the Royal Astronomical Society,
and British and Scottish Meteorological
Societies, etc.

1882.

INTRODUCTORY REMARKS.

EVERYONE who is at all acquainted with the history of practical astronomy must be familiar with the name of Hevelius. He was born at Dantzic in the year 1611, and at an early age became attached to the study of mathematics and astronomy. He published a variety of astronomical works, the chief of which were " Selenographia," " Cometographia," " Mercurius in Sole Visus," "Machina Celestis," "Annus Climactericus," and "Podromus Astronomiæ." This last work, containing a catalogue of 1,564 stars, was not published till 1690, about three years after his death. A short account of Hevelius and his controversy with Dr. Hooke will be found in the preface to his large catalogue of stars, by Francis Bailey, Esq., in the thirteenth volume of the " Memoirs of the Royal Astronomical Society." In reference to this controversy I have found a manuscript note, by Miss Caroline Herschel, which I have thought of sufficient interest to reproduce here. It was pasted upon a blank leaf of the first volume of a copy of the " Machina Celestis," which is the work immediately under consideration, and dated August 29th, 1799. It will be seen that Miss Herschel has inadvertently called him Rooke instead of Hooke.

" *In this first part Hevelius has described all the instruments which were superior to Tycho Brahe's, made use of in his observations.*

" *The second part (Gedani 1679 in Folio reg: 14 alphabetical 6 sheets with many copper plates) contains his observations of 48 years. It is much to be lamented that most of the copies were in the same year destroyed by fire.*

" *Robert Rooke doubted the consequence of these observations ; but the celebrated Halley having been sent by the Royal Society to Danzig, found them very correct, and approved of all the author's methods. The famous astronomer Is. Bullialdy has also examined Hevelius's observations, and the opinion of both these lerned men may be seen in the letters which are given under the titul Excerta ex &c: &c:*"

B

Upon all the works published by Hevelius he spared neither time, money, nor expense in their typography and illustration. The engravings relating to the three selected chapters are proof of this; and having first photographed them, I have had them reproduced by a permanently printing process.

In copying these plates I have very much reduced their size, and I think they may be considered to be interesting as illustrating the precise method which he adopted in taking his observations, as well as his attitude and personal appearance.

Although, as I have just stated, many persons are familiar with the name of Hevelius, yet it is highly improbable, considering the extreme scarcity of this his principal publication, that many persons are aware of the indefatigable zeal, perseverance, and ingenuity with which he pursued his favourite science. It would appear that nearly all the mechanical arrangements for the construction of his instruments were effected under his direction and personal superintendence, and he displayed therein no little taste in the ornamentation of his various sextants, quadrants, and their several appendages. In his "Machina Celestis," the work above referred to, he has given a very full account of his optical instruments and mechanical appliances, and it is a matter for sincere regret that nearly the whole impression of this very valuable work was destroyed by fire almost immediately after its completion. The few copies which escaped the general conflagration of his house, books, and instruments were those which he had just distributed to a few personal friends. In the preface to his "Annus Climactericus," the last work which he lived to publish, Hevelius thus alludes to his irreparable loss :—" Nam, cum die 26 Septembris " ejusdem dicti anni, immane et atrocissimum illud infortunium, " ex proposito et summâ malitiâ bipedum nequissimi cujusdam " hominis, et quidem mei famuli mihi acciderit, ut nefandis " ignibus (id quod privato cuidam sicuti mihi, câ ratione, " quantum sciam, haud unquam fortè obtigit) ædes meas " numero septem, cum omnibus meis in iis existentibus rebus, " nummis, auro, argento, omnique prorsùs suspellectile, typo- " graphiis, magnâ parte bibliothecæ, omnibusque operibus, ab

"anno 1647 ad annum 1679 sumptibus meis editis, cum
"pretiosissimâ meâ Uraniâ, speculâ, cunctisque organis, tàm
"astronomicis, quàm opticis, in Machinæ meæ Celestis parte
"priori descriptis ac delineatis, alia quàmplurima pretiosa, ut
"modò taceam, penitùs perdiderim; atque ita unius, aut
"alterius horulæ spatio, omnibus ferè meis substantiis, et facul-
"tatibus exutus omninò fuerim, facilè, mi lector, colligere datur,
"quomodo talis inspiratus, subitaneus, horrendus ac tristissimus
"casus, aliquem etiam longè animosiorum prosternere, circu-
"losque ejus omnimodè conturbare valeat. Sic ut mirum haud
"fuisset, me, ex immani istâ consternatione, ac subsecutis
"curis, sollicitudinibus, et afflictionibus, imprimis dum mihi in
"memoriam revocarem quænam possederim, quænam perdi-
"derim, quorum plurima nunquam reparare, ac recuperare
"valerem, in illo ipso momento ad operas omnes incapacem
"prorsùs redditum fuisse, imò ipsam vitam derepentè cum
"morte commutasse."

Notwithstanding this terrible misfortune, Hevelius summoned
up courage to build another observatory, which, however, was
less complete than his former one, and he continued, during
the few remaining years of his life, some observations upon the
planets, as well as those comets which appeared during the
interval. So far as I know, there are only three copies of his
"Machina Celestis" in this country, and notwithstanding the
repeated enquiries which have been made of the chief conti-
nental booksellers, not a single copy has been obtainable for
many years. Under these circumstances I have thought that a
reprint and translation* of a certain portion of the first volume,
which treats more immediately of the method which he adopted
with respect to the mounting and equipment of his telescopes,
was desirable. For this purpose I have selected Chapters xviii.,
xxi., and xxii. It must not be supposed that by a reprint of the
aforesaid chapters I anticipate some important practical information will be elicited, but merely that it will be the means of hand-

* For this translation I have to express my thanks to Frederic Hale Forshall,
Esq., of Upperton Road, Eastbourne, who, hearing of my intention to give one,
very kindly relieved me of the responsibility—scientiæ causâ.

4

ing down, for some future years, the ideas which were entertained upon this subject shortly after the invention of the telescope, as well as to show the endeavours which were made by Hevelius to turn Galileo's grand discovery to the best account, and at the same time to afford both instruction and gratification to those persons who had become excited, in no small degree, by the gleams of information which had reached them from the few who had seen, albeit with very imperfect means, the newly-discovered astronomical wonders.

C. L. PRINCE.

The Observatory,
Crowborough, Sussex,
August 4th, 1882.

PLATE I.

PLATE 2.

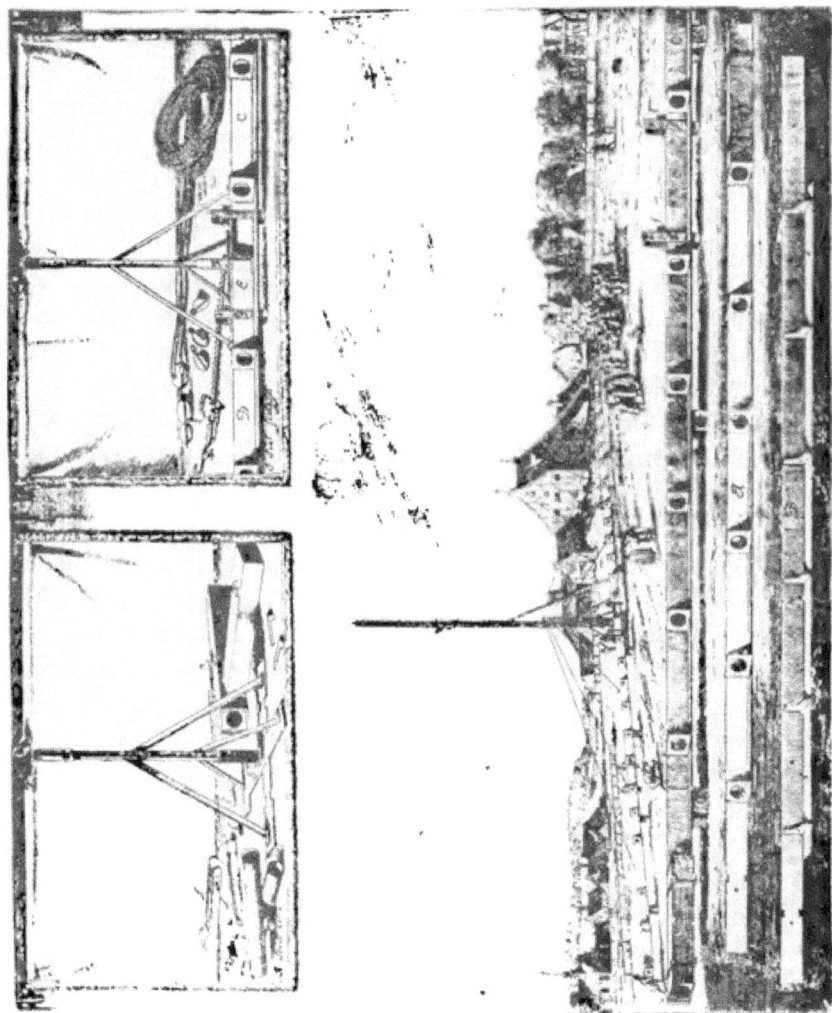

PLATE 3.

MACHINA CELESTIS.

CAPUT XVIII.

*De Peculiari Machinulâ Helioscopicâ, pro Eclipsibus, Maculisq̃
Solaribus exquisitè determinandis.*

MAchinæ illius Helioscopicæ beneficio, jam olim à nobis detec-
tæ, ac etiam pag. 98 Selenographiæ satis plenè descriptæ, Maculas, Eclipsesque Solares multò, quàm hactenus, præci-
siùs observari ac delineari posse, ut ipsa experientia docuit, sic
omnes puto spontè concessuros. Cùm autem totus Machinæ ejus
Helioscopicæ apparatus, utpote Globus ille in fenestellâ volubilis,
sive versatilis cum scamno, Tabellâ, omnibusque iis, quæ his
appertinent, satis dilucidè ibidem sint exposita, nolo hoc loco
de iis plura facere verba, sed Lectorem eò remitto. Veruntamen,
quoniam dicta illa Machina, in Selenographiâ nostrâ delineata,
fundamentum quasi est hujus recentis adumbrandæ, operæ
pretium esse duco, ejusdem saltem typum, hic apponere; atque
sic res ipsa non solùm multò levior erit mihi descriptu, sed
universa recentis hujus Machinæ intellectu quoque Tibi erunt
longè faciliora.

Habes itaq̃; priorem illam Machinam Helioscopicam,
Benevole Lector, in adjuncto Iconismo accuratè delineatam;
non dubito, quin Siderum Cultores haud magno negotio, quà
nimirùm ratione Maculæ, Eclipsesque Solares animadverti
debeant, omnia intelligant: & si quæ fortè minùs pateant, ex
Selenographiæ nostræ Prolegomenis ea percipere erit proclive.
Adeò, ut hæc Delineatio jam affatim plurimas partes alioqui
describendas ostendat, quibus etiamnum adhuc utimur: inter
cætera verò idem adhuc Globus versatilis, Scamnum, quod duo
statumina perpendiculariter erecta spiratim striata habet, cum
duabus suis pericochleis, quarum beneficio asser iste, super quem
tabella est imposita, atque attolli & deprimi pro quâvis Solis
Altitudine potest, mihi sunt in usu; pariter Telescopium, canali

inditum, atque ope alicujus vectis quadriformis tabellæ orthogonaliter adaptatum : quò Sol rectà per radios scilicet rectos in
Tabellam incidere, discumque suum in chartá depingere pòssit.
Ita ut hucusque hanc observandi methodum nondum correc-
tiorem reddiderim, exceptá unicâ solâ parte, nimirùm ratione
tabellæ *b*, in quá Solis discus excipitur, atque charta, circulum
sive discum *f rv u z* gerens, annectitur.

Hæc, inquam, Machina utique est modus, prout exploratum
habemus, ad Maculas, Eclipsesque Solares excipiendas & deter-
minandas exoptatissimus ; nisi quòd Sol ratione ejus motûs
velocissimi, præsertim diurni, ob quē non solùm assiduè ascendit
vel descendit, sed etiam ab occasu ortum versùs commovetur
(ut more Peripatetico loquar) ac in perpetuo versatur motu, ut
nunquam in tabella, vel potiùs circello ne quidem per temporis
momentū quietè subsistat ; adeoq ; si aliquid accurati deprehēdi
debeat, necessum omninò habeas, Tabellam in continuâ agita-
tione detinere, globumq ; versatilem aliter atq ; aliter dirigere,
nunc magis magisque attollere aut demittere : quo in opere nisi
aliquis exercitatissimus Tibi sit ab auxiliis & manibus, profectò
arenam metiris, præsertim, si solus & Machinam perpetim com-
movendam, & phænomena simul designanda habeas.

Incommodum itaq ; hocce probè perspiciens, cumprimis anno
1661 in Eclipsi Solari, præsente Eximio Bullialdo, Hospite tunc
longè exoptatissimo annixus sum, quâ ratione difficultates istas
convenienter prorsùs tollerem, atq ; rem eò deducerem, quò Sol
in objectá tabellá, circelloque ejus descripto quasi immotus
subsistere possit, vel quod eodem recidit, quomodo Sol in circello
assiduè immotus, ad Maculas, phasesque Deliquiorum exquisitissimè absque omni titubatione designandas, detineretur.
Ad quod propositum exequendum Clariss. Bullialdus me eo
tempore, ut quantocyùs, omnium Observatorum bono, id per-
ficerem, haud parùm excitabat. Videbatur quidem nonnemini,
ut ut alioqui Mathematicarum Rerum satis gnaro, non tantùm
res magnæ difficultatis, sed etiam ad ipsam praxin commodè
deducenda penè impossibilis. At mihi verò non adeò absona
res visa est ; quippe non ignorans, quòd quemadmodùm jam in
Pinnacidiis **Quadrantum** Solem firmiter, ne circello unquam

exiret, detinere in mea esset potestate, me quoque, Divino auxilio adjutum, id sine omni dubio feliciter assequuturum: sicuti etiam brevi pòst ex voto nostro successit negotium.

Totum autem artificium nonnisi in Tabellà illà, Globo versatili sive Telescopio affixà, latet. Prior namque Tabella illius Machinulæ Helioscopicæ, in Selenographià descriptæ, scamno solummodò innixa, secundùm eandem semper Solis longitudinem commoveri datur, nisi asserem, in pericochleis insidentem, earum beneficio attollas vel deprimas. Id quod autem, simul commovendo Tabellam, adeò velociter, ut motum Solis prorsùs assequaris, vix unquam conceditur. At verò Tabella hæc recèns inventa, quà nunc utor, licet scamnum in eàdem semper altitudine, eodemq; situ retineatur, nihilominùs regimine duarum cochlearum minimarum ad Solis motum omnimodè & exactissimè dirigitur, ut Sol perpetuò in circello subsistere, ac ne punctum quidem eo discedere appareat. Dicta autem Tabella orbicularis *g* ex diversis componitur partibus, quas, quoad fieri licebit, melioris intellectùs gratià singillatim describemus; quæ verò adeò dilucidè haud poterunt, ex apposità singularum partium, distinctè satis adumbratarum Delineatione elucescent, vel si mavis potiùs ingenio tuo, & manu operi adhibità facilè comprehendes.

Primò opus omninò erat, ut Tabella *g*, in quà charta cum suo circello observatorio affigi debet, quàm lenissimè, simul velocissimè ad motum Solis Azimuthalem, juxta motum scilicet rectum, horsum prorsum commoveretur. Secundò, non minùs quòque summoperè necessum erat, ut simul eadem Tabella cum suo circello, in eàdem descripto, secundùm motum circularem, nimirùm in Orbem, æque promptè ac citissimè circumagi posset. Ad quod aptè perficiendum duo peculiares Orbiculi lignei diversæ magnitudinis, meo judicio, requiruntur, ita ut major *g* minori *c* beneficio axiculi *m* conjungantur, si videlicet duplex iste motus rectilineus & circularis unà & simul obtineri debeat.

Arripui itaque duos asserculos tiliaceos levissimos, orthogonaliter conglutinatos, longitudine 16 ferè pollicum, prout ad *a* vides. Ad superiorem erecti asserculi partem lamella, adjecta cochleà perpetuà longiori *b*, transiente per duas fibulas, aliis

quibusdam minoribus cochleis est adstricta.   Dein, per Orbiculum minorem *c* columellam ferream *m* trajeci, adaptato ei à parte
posticâ Orbiculi *c* axiculo striato *i*; atque sic columellam istam
m, unâ cum toto Orbiculo *c* per foramen *k* asserculi erecti *a*
transversam, à tergo illius asserculi pericochleâ firmiter ad-
strinxi.   Quibus partibus sic inter se conjunctis, atque cochleâ
perenni *b* conversâ (cùm hujus dentes strias axiculi *i* perpetuo
motu mordeant) omninò necesse est, ut totus hicce Orbiculus
minor *c* se in orbem, ad convolutionem & gyrationem cochleæ
perpetuæ, huc vel illuc pressæ, volvat.   Dehinc, in latere Orbiculi minoris *c* antico peculiare Instrumentulum *d* affixi, & quidem
prorsùs simile illi, quod Capite V pag. 115 ad Quadrantē Hori-
zontalem, vel potiùs ejus columnam, cui Quadrans iste tum
affigebatur, olim applicaveram; quò ad æquilibrium, vel libellam
dirigi & rectificari posset.   Quod autem cùm ibidem abundè sit
descriptum, nolo eidem diutiùs immorari; constat alioquin ex
unicâ folâ cochleâ *c*, partem intermediam robustiorem horsum
prorsum commovente.   In quâ parte intermediâ *d*, cui annexa
est dicta cochlea *c*, duo sustentacula *ff*, ad angulum normalem
incurvata, sunt conferruminata; eo quidem fine, quò major
tabella orbicularis *g* hisce binis sustentaculis *ff* imponi, coch-
leisque minusculis affigi validè queat.   Atque tum, duobus illis
Orbiculis *c* & *g* combinatis, integrum cuiq; erit, convolutione
cochleæ *e* majorem Orbem horsum prorsum ad motum
rectum ex voto commovere; directione verò alterius cochleæ
perpetuæ *b* utrumque Orbiculum, tam *c*, quàm *g* conjunctim in
orbem circumducere, & quidem adeò leniter ac promptè, ut vix
quicquam exquisitius ac velocius unquam possit.

   Atque hæc tota structura est hujus recentis Machinæ Helioscopicæ, cujus ductu omnem Solis motum exactissimè, tam
secundùm longitudinem, quàm altitudinem assequi & concomi-
tari nullo ferè negotio Tibi erit integrum; sic ut Sol hâc ratione
omninò immotus in circello Observatorio, ut suprà dicebam,
subsistere cogatur, & ne latum quidem unguem ex eo discedere,
prout mox pleniùs docebitur.   Necesse autem est, ut totam hanc
Machinulam baculo, sive vecti quadrato *b* applices, immittendo
scilicet vectem per foramen quadratum *a*, ita ut ad ductum

canalis, sive Telescopii Globi volubilis inserti, ad angulum omninò normalem dicta Machina cuneis adstricta firmissimè quiescat.
Deinde, tota ista Machinula cum binis suis affixis Orbiculis,
minori c & majori g, in quo charta cum suo descripto circello
Observatorio claviculis est annexa, scamno Helioscopico, prout
ex Figurâ W elucet, imponatur, atque rudiori tantùm Minervâ
ope pericochlearum huc illuc utrumque Orbiculum commovendo,
quò radii Solares circellum n penè impleant, ad Solem dirigatur.
Tum demum scalam à parte posticà Socium ascendere jubeo;
qui, arreptâ sinistrâ manu alterâ cochleâ c, tabellam majorem g
unà cum circulo Observatorio secundùm motum directum diri-
gente; dextrâ verò alterâ cochleâ b, minorem Orbem c unà cum
majori g in gyrum commovente, poterit hanc vel illam cochleam
huc vel illuc torquendo mirè faciliter omnem Solis motum ita
exquisitissimè & lenissimè eximere, ut Sol ipse ne minimùm
quidem limites circuli Observatorii n exire, sed quasi in circello
immotus sive affixus videatur. Is igitur, qui præcipuo Observa-
tionum Directori adstat, Totâ Observatione nihil quicquam ampliùs agit, quàm ut assiduè ad discum Solis, circellumque Observatorium fixis oculis assiduè attendat, ne radii Solares nullâ
ullâ parte limites egrediantur, sed perpetuò inclusi & circumsepti
penitùs permaneant: quemadmodùm id etiam cuilibet proclive
admodùm factu est, nec non abundè ex superiori Figuratione intelligitur. Atque hâc ratione cuilibet Observatori erit expeditum,
tam Maculas Solares, quàm ipsas Eclipses promptissimè, cor-
rectissimè, absque omni directionis molestiâ ac motus Solaris
impedimento summâ cum jucunditate, ludendo quasi, & quidem
multò scrupulosiùs, quàm hactenus unquam, designare, ac toties,
quoties lubet, eandem Observationem resumere ac corrigere.
Adumbratis autem sic omnibus, necesse quoque est, simul in
limbo superioris circelli observatorii punctum verticale annotare,
ad inclinationem Solis beneficio funependuli, ad unum serratum
ex fibulis mobilem appensi, obtinendam; prout jam in Selenographiâ nostrâ suo loco plenè commonstravimus.
Ne verò tota illa Machinula Helioscopica cum omni suo
apparatu ferreo se se supinet, vel vectem b, simulque ipsum Telescopium, Globo insertum, nimio suo pondere vel incurvet, vel

distorqueat, res flagitat, ut Coadjutor, cochlearum scilicet moderator statim initio totam Machinulam à tergo phalangâ quâdam *o* suffulciat, immittendo scilicet alteram ejus extremitatem ad pavimentum inclinatè constituendo, sicque totum Instrumentum securè in vero suo situ, quoad volueris, conservabis.

De cætero verò notes velim, cùm Orbiculus *g* satis sit amplus (ut ut sufficiat, diametrum ejus integrum tantùm pedem exæquare) quòd in hâc recentiori Machinulâ Helioscopicâ duplici vecte *b* & *l* opus habeas, alterâ scilicet breviori *l*, & alterâ longiori *b*, sicuti distinctè admodùm ex Icone patet, ut res fusiori declaratione haud indigeat. Et licet duplici hoc vecte Instrumentulum regatur, minimè tamen inde vel quicquam difficilior evadit directio, aut ipsa Observatio, quàm illo uno simplici vecte, sed æq̃ ; facilè omnia peraguntur ; commotis enim binis illis cochleis, tam Orbiculi, quàm Globus versatilis cum suo tubo cedunt, ac se se ad nutum commovent.

Hocce artificium quàm aptum & exquisitum sit, ad denotanda imprimis Solis Deliquia, non meum est hic prolixè celebrare, sed alii, istud aliquando ad Observationes adhibituri, sine dubio, abundè agnoscent. Certo certior sanè sum, hanc methodum nemini hucusque adhuc (sit venia verbo) fuisse cognitam, multo minùs usitatam : ideoque omnibus Astrosophis, quibus Observationes nunquam in tam sublimi negotio nimis exquisitæ esse possunt, eam haud usque adeò fore ingratam spero, præprimis, si quævis rectè perceperint, ac singula ad typum & descriptionem hanc nostram, vel etiam in Selenographiâ nostrâ jam traditam, exstrui curaverint. Abrumpo igitur, me accingens ad Telescopia, Tubosque Opticos ; quippe qui, velut ex dicendis abundè patebit, egregium suum usum in Rebus Astronomicis detegendis & contemplandis præbeant.

CAPUT XXI.

De Telescopio nostro maximo.

Cùm abundè nunc tandem exploratum sit, quantum plurimis Observationibus Cœlestibus, Rei Literariæ bono instituendis contribuant præ aliis exquisita & perfecta Telescopia. Hincq; nemo non Astrophilorum ac Opticorum nihil habet potius, quàm ut indies perfectiora, insuper & longiora conficiat, construatque. Longiora tamen sexaginta vel septuaginta ped., quod sciam, ut ut lentes pro longioribus Tubis hinc inde fortè expoliverint, feliciter, & cum aliquo successu ad Sidera hactenus vix ullibi exhibita fuerunt. Eâ de re, quòd hucusque nondum convenientem Machinam, Tubumque pro inferendis lentibus adinvenire & componere potuerint, & quidem talem, qui omni commotioni & directioni expeditè pareret, ac Observatoribus minimè esset mole suâ molestus. Imò plurimi Eruditorum, hoc ipsum adeò faciliori viâ detectum iri, penè desperarunt ; nihilo tamen seciùs spem omnem minimè abjeci, sed amplissimam ejusmodi Machinam, Favente Deo, pro ferendis lentibus centum, imò centum quinquaginta pedum utique elaborari, atque ad usum commodum transferri posse, planè sum confisus. Ad quod opus suscipiendum imprimis Illustr. Dn. Titus Livius Burattinus, ut Rerum Mechanicarum omnium, sic & harum Opticarum Peritissimus, & Exercitatissimus me maximè inflammavit : cùm non solùm lentes, manu suâ expolitas, pro Telescopio centum quadraginta ped. promptissimè promiserit, sed etiam brevi pòst liberali animo exhibuerit. Initiò tamen, rem istam plurimas habere difficultates, optimè me prævidisse fateor. Etinim, si negotium aggrederer superiori viâ, construendo videlicet canalem quadratum ex quatuor asseribus, atq; ex quatuor partibus conjungendis, quamlibet licet quadraginta ped. longam. Istæ quidem partes singulæ, largior, tam anti detecto, quàm alio, adhuc diverso modo coadunari, atque beneficio quarundam cistarum ac funium ad rectitudinem deduci possent ; si quis nimirùm haud vulgares sumptus, nec singularem potentiam ad

Machinam componendam & regendam non vereretur. Verùm sedecim asseres quadraginta ped. longit. sufficientis crassitiei & latitudinis, unà cum tribus adeò robustissimis cistis, quindecim minimùm ped. longis, tanto ferro munitis, nec non tot funibus, trochleis superadditis, tantòque apparatu, quantæ, quæso, essent molis, quanti laboris & operæ, ea omnia conjungere, ad lineam rectam deducere, elevare, commovere, atque in suo semper situ & rectitudine conservare?

Idcircò prævidens hanc insignem molestiam, negotium hocce aliâ planè singulari ratione tentavi, nimirum, ut ejusmodi Machinam pro lentibus centum quadraginta ped. ex octo tantùm, & quidem multò arctioribus & subtilioribus asseribus abiegnis construerem; quæ Machina ut esset sic levior, ita pariter ad construendum & dirigendum longè commodior ac facilior, præprimis cùm omnibus cistis intermediis prorsus remotis, imò, quod maximum, absque omni ferro, ne unico quidem clavo ferreo unius obuli valoris (quod notandum) adhibito componeretur : & ut rem brevibus complectar, operam dedi, ut Machinam ex solis duobus asseribus gracilioribus, rotundum quasi Tubum præsentantem conficerem. Id quod principiò quibusdam, ut ut alioqui non omninò rudioribus, ferè absonum, imò ad instar paradoxi videbatur; canalem ex duobus solummodò planissimis asseribus exædificare & instruere. Quandoquidem extra controversiam penitùs est, duos asseres explanatos nunquam eâ ratione, ut tubum clausum referant, lucemque ab omni parte excludant, ita combinari ac conjungi posse. Nihilominùs feliciter, cum Deo, successit opus, ac desideratum finem obtinuerim, ac instar veri tubi eo commodè uti possim : quanquam, ut facilè intelligis, hanc ipsam Machinam construere, tum regere ac dirigere, res sit alicujus difficultatis.

Machinam autem ipsam ex quatuor asseribus, quadraginta ped. longis, quò debitam longitudinem centum quinquaginta circ. ped. obtinerem, composui. Primò assamentum decem vel undecim poll. ferè latum, ac uno & quartâ ejus parte crassum, æquabiliter explanatū & extensum arrectis lateribus super aliud assamentum, tantummodò octo vel novem poll. latum, & vix tribus quartis unius digiti crassum ad medietatem penè explanati ejus

lateris, tanquam ad basin, orthogonaliter erexi & affixi : prout ex adjuncto Inconismo ad *a* B *a* videre est. Et ut eò firmiùs eâ ratione conjuncta hæc duo assamenta consisterent, plurimis anteridibus & tenaculis ligneis *c* secundùm longitudinem impositis, ad tres vel quatuor pedes à se invicem remotis, uno scilicet latere in basin, asserem scilicet inferiorem *b*, altero verò latere anteridum *c* admovendo alteri assamento orthogonaliter arrecto, & quidem à latere illius asseris arrecti postico, ibidemque ligneis clavis firmissimè constrictis & conglutinatis statuminavi & corroboravi; ne vel minimùm hi bini asseres, tam subjacens, quàm arrectus, alter ab altero commoveri, vel se se distorquere possent, sed perpetuò eo in situ conjunctim conservarentur. Nam, quemadmodum jam Cap. VIII occasione illius Quadrantis meminimus, assamentum illud *a* B *a* ad latitudinē super alterum explanatum *b* *b* erectum, nullo modo sinit inferius, basin videlicet subjectam *b* *b*, se se incurvare, licet assamentum alioqui satis sit arctum & gracilentum; è contrario hocce assamentum subjacens *b* *b*, cùm suâ planitie arrecto validè sit affixum, neutiquam permittit, ut superius *a* *a* arrectum ullo pacto se se obliquare queat. Ita ut hæc bina assamenta quadraginta ped. longa dictâ ratione inter se firmissimè combinata & conjugata, semper strictissimè in rectâ omninò lineâ se se conservare, & nullo modo inflectere possint, imò citiùs, maximâ vi addibitâ, asseres illos ita compositos franges, quàm obliquabis : quemadmodùm facile intelligitur.

Ab altero autem latere illius arrecti asseris *a* *a*, ut ad A & D manifestum est, è regione anteridum & tenaculorum *c* asserculos quadratos *d*, sed amplo foramine præditos unius ferè pedis magnitudine insuper orthogonaliter erexi, & ad utrumque assamentum, tam subjectum *b*, quàm arrectum *a*, ligneis clavis probè affixi, atque conglutinavi; eo non solùm fine, ut utrumque assamentum magis magisque constringant ac corroborent, sed in primis, reliqua ut modò taceam, quò lumen adventitium, ab exteriùs proveniens, omninò prohibeatur, ne in lentes incidat, oculosque Observatorum irradiet. Combinatis itaque ac compositis quatuor illis partibus *ABCD*, artis nunc est, eas invicem fecundùm longitudinem conjungere & constringere, ut ne minimum quidem ferri res opus habeat, & nihilominus debitam suam habeat firmitudinem, etiam illicò rursùs haud multo labore, temporisùque dispendio restringi & resolvi possit. Ad hoc ipsum autem præstandum commodius remedium nullum planè invenire potui, quàm ut peculiares anginas ligneas constringentes sive constrictorias conficerem, quibus alioqui Scriniarii utuntur, dum asseres conjungunt & conglutinant, quales in præcedenti Figurâ, & superiori Tabellâ ad sinistram ad *f* delineatas dedi; hoc tandùm discrimine, quòd Scriniariorum anginæ ab utrâque extremitate firmiter clausæ ac conglutinatæ sint, & quicquid in iis constringendum & arctandum est, intra aperturam, sive inter utrumùque crus inferendum sit; hæ verò nostræ anginæ constrictoriæ, pro componendâ Machinâ hâc nostrâ, tantùm ab unâ extremitate sint clausæ & conglutinatæ, ab alterâ verò extremitate peculiari ligillo quadrato, subscudibus sculpto (loco securiculæ) per strias anginæ utriusùque lateris immisso occludantur, trajecto clavo ligneo, ne hocce ligillum exeat, sitùque ad resistendum eò potentius, tum omni tempore illicò, quando opus, rursùs eximi possit. In binas has anginas *f* duorum arrectorum asserum *D* & *C* extremitates conjunctim, sed exemptis primùm ligillis illis *e* ex anginis, à parte superiori in apertas anginas strictè immittuntur, cùm fissura anginæ omninò crassitici utriusù; asseris arrecti respondeat; vel, quod præstat, binæ hæ anginæ *f* per foramina adæquata subjecti asseris *b*, baseos scilicet Machinæ, à parte inferiori sursum depanguntur: deinde dictæ anginæ ligillis illis quadratis *e*, ne ullo modo exilire vel rumpi queant, clauduntur, clavisùque suis obfirmantur. Posteà duos cuneos robustissimos ligneos, ad anginæ latitudinem planè formatos, sed sibi invicem obversos, in utramùque anginam, inter securiculam *e* nimirùm & binos asseres conjunctos immitto & depango, eosùque paviculâ cousùque validissimè cogo, donec ampliùs cedere recusant, atùque utrumùque assamentum *D* & *C* satis superùque sit constrictum & coarctatum: prout ad *E*, inprimis ex superiori tabellâ ad dextram appositâ, videre est. Atù; ita duæ hujus Machinæ partes debitè sunt conjunctæ, ut nullà vi, nisi cuneos recludas ac relaxes, disjungi

& separari queant. Pari nunc ratione reliquas etiam *A* & *B* similibus quatuor anginis, cuneisque ad invicem constringuntur, ut omnes quatuor partes *A B C D* ordine secundùm lineam rectam beneficio trium parium anginarum cohæreant.

Probè autem notandum, quòd in Machinæ medio ad *E* statumen aliquod *F*, super basin quandam decussatam *G* erectum, asseribus orthogonaliter arrectis imponendum & insternendum sit, anginisque amplectendum, ac simul cuneis illis angendum, quò ibidem Machinæ validè inhæreat, & nullo modo ab eâ discedat. Estque statumen dictum ad duodecim pedes altum, atque duobus anteridibus & retinaculis benè corroboratum : quemadmodum ex Delineatione superiori liquet. In statuminis vertice bini orbiculi super axe volubiles ad *i* & *k* inserti sunt ; quo sine verò deinceps exponetur. Dispositis itaque his omnibus, ac partibus hujus Machinæ firmè inter se conjunctis, super imposito illo statumine decussatâ suâ basi ; adhæc, totâ Machinâ ad arborem, quâ elevari debet, deductâ, ibidemque ad libellam super truncos, sive sedilia omninò in directum protensâ, curvaturisque omnibus Machinæ exemptis (quò per singula foramina *d* ab unâ extremitate ad alteram liberrimè pateat prospectus) diversissimi rudentes, funesque, nec non trochleæ plurimæ diversi generis Machinæ applicantur & adstringuntur : quemadmodum ex superiori Schemate elucet, in primis ex hâc appositâ adhuc paullò dilucidiùs, ubi Machina jam sublata pendet.

Verùm plurimùm interest, quo loco trochleæ instringantur, ut æquipondium videlicet quâ longitudinem invenias, atque funes gestatarii, sive portatarii æquali potentia Machinam arripiant, æqualiterque ubiq ; sustineant. Nam, si binæ illæ trochleæ intermediæ *b* & *c* ad columnam nimis vicinæ ratione statuminis *f* annexæ fuerint, Machina terram versùs utràque extremitate se se incurvabit, convexitatem scilicet circa medium sursum exhibendo ; contrarium verò experieris, si trochleas illas binas *b* & *c* longiùs, quàm oportet, alligaveris, extremitates se se erigent, totaque Machina terram versùs contrahet convexitatem. Adeò, ut artis planè sit, initiò, quando Machina primùm adornatur, trochleas non solùm *b* & *c*, sed & reliquas *a* & *d* ita disponere & applicare, ut pari potentiâ ubique funes gestatarii

totam Machinam sustineant, atque in rectâ lineâ sustentent.
Huic rei autem maximum momentum accresceet, si rectè noveris
ac exploraveris, à quo arrectorum asserum latere, an antico, an
verò postico trochleæ sint annectendæ, vel quænam ab uno, &
quænam vicissim ab altero adstringendæ. Etenim, nisi hæc omnia
probè explorata habeas, nullâ ratione totam Machinam ad libellam,
ut universi asserculi quadrati ac perforati *d* arrecti ad ductum
Horizontis parallelum consistant, deduces, sed varias curvaturas
ratione inclinationis deprehendes ; quanquam parùm intersit, an
Machina hæc paullulum se se convertat & inclinet, dummodò
rectam omninò lineam quà longitudinem perpetuò exhibeat.
Præstat tamen, totam Machinam non tantùm prorsùs in
directum protensam, sed etiam ubique ad perpendiculum erectam
habere. Isthoc pacto namque quævis objecta proclivius reperies,
quàm si Machina transverso & acclinato aliquo situ pendeat.
Suasor itaque sum, ut omnem adhibeas industriam, quò
Machinam dictam primùm tum quà longitudinem, tum quà
inclinationem rite & exactissimè constituas ; id quod utique fieri
etiam poterit, si solummodò laborem non verearis.

Quomodò verò, & ubinam trochleæ omnes secundùm pro-
portionem exhibitæ nostræ Machinæ sint alligatæ, promptiùs ex
Delineatione ipsâ, quàm prolixâ informatione addisces. Primò
quatuor trochleas ad asseres erectos anneximus ; nimirùm
priorem ad *a* antecedentes anginas, alteram ad *b*, tertiam ad *c*,
& quartâ ad *d*, per quarum orbiculos rudens fortissimus, aliàs
gestatarius nobis dictus, trajectus est, eâ ratione, ut alteram
rudentis extremitatem primùm per *d*, dein *c*, rursùs per orbiculum
inferiorem duplicis trochleæ *e*, hinc per trochleam *b* & *a*, &
deniq ; iterum per duplicis illius trochleæ orbiculum, sed
superiorem trajecerim, extremitatibus ad *f* constrictis. Rudentis
autem longitudo necesse sit omninò operi adæquata : quandoquidem, si nimis brevis est, nullo pacto satis validè sustentat
Machinam ; rursùs si debito longior est, ut ultra statumen *k*
exporrigatur, ægrè admodùm Tubus in suo arrecto situ conser-
vatur : quippe statumen *F* facilè infra trochleam duplicatam *e*
transiret, ac se se converteret, maximo Machinæ dispendio ;
quòd si verò debitam hujus rudentis longitudinem observaveris,

statumen *F* potest se se ad rudentem reclinare, vel quod idem, rudens, per duplicem trochleam *e* duplicatò trajectus, statumen *F* sustentat, ne se se invertat, atque ab arbore discedat, sed situm suum eò meliùs conservet. Sublatâ itaque duplici hâc trochleâ *e* ope funis ductarii *y x* in altum, protinùs rudens gestatarius *d c e b a e f* ubique in suis locis æquabiliter prorsùs Machinam apprehendit & sublevat : quippe dictus rudens liberrimè per omnes orbiculos incedit, & nullibi remoratur, imò pro diversâ elevatione promptissimè duplici trochleæ *e* cedit; sicuti Mechanicarum Rerum periti facilè intelligunt. Quòd si verò duos diversos funes, alterum scilicet ad *b* & *c*, alterum rursùs ad *a* & *d* alligassem, eosque tantùm per binos illos orbiculos *e* duxissem, fieri sanè haud potuisset, ut Machina adeò æquabiliter tolleretur, ac in lineâ rectâ confirmaretur. Mox enim hic, mox ille funis laxior, nunc strictior foret, quod illicò Machinæ obliquitatem procrearet. Nam scias velim, Machinam hanc vastissimam, ob nimiam suam longitudinem, ut ut partium compactio & constructio sit penitùs validissima, procliviter tamen admodùm curvaturam inducere, atque à lineâ rectâ discedere; hinc nisi huic incommodo subvenias, Machinam nimirùm ope funium, per plures trochleas sibi invicem cedentium, æquabilissimè ab omni parte attollendo, profectò operam egregiè ludes, atque perdes.

Verùm enimverò, unicus hicce rudens, per sex illos orbiculos trajectus, licet ferè sit omnium primarius, neutiquam tamen adhuc folus huic rei sufficit; quò videlicet tota hæc Machina in debitâ suâ rectitudine conservetur. Etenim, licet circa Machinæ, medictatem ad *a b c d* officium suum præstet, neutiquam tamen ipsas extremitates *g* & *i*, ut istæ se haud evidentissimè Horizontem versùs inflectant, constringere valet. Proindequò & huic rei succurrerem, volui primùm statumen illud *F*, in basi suâ *G* erectum, Machinæ circa medium imponere, ibidem que duplici anginâ constringere, aliisque tenaculis renitentibus obfirmare, ne se se facilè perverteret; quâ verò ratione hæc omnia sint confecta & combinata, nec Schemate adeò dilucidè commonstrare, nec verbis, ut chartæ parcerem, describere potui; oportet igitur, sis, ut ipse reliqua inquiras. Super istias nunc statuminis orbi-

culum *k* superiorem pariter, ut in præcedenti Tubo sexaginta
ped. factum est, alium rudentem satis crassum (quem directarium
appellare lubet) extremitatem Machinæ *g* versûs duxi, ibidemque, ubi nempe moles id requirit, extremitatem illius rudentis
annexi, vel annexo unco ibidem appendi; alteram verò rudentis
illius extremitatem *b* ad Ergatam (quæ vecte dentato trium circiter pedum longitudine, ac striato axiculo, alioque artificio in-
trinseco constructa est) ad *i*, circa Machinæ extremitatem alteram
inferiorem adstrictam, validissimè alligavi; sed oportet, ut rudens iste directarius in antecessum, si recens est, optimè sit
distentus, ne se se nimiùm deinceps dilatare & expandere possit.
Quandoquidem hicce rudens *g y h* eam ob causam adjectus est,
quò liceat utramq; Machinæ extremitatem superiorem & inferiorem regere, tantumque elevare, ut resurgant, quantum à rectitudine Machinæ deliciant: quemadmodùm quoque factu ad-
modum est proclive. Nam dum Ergatam, vel ejus vectem serratum *h i*, qui loco cocheæ perpetuæ servit, manubrii beneficio
circumagis (cujus faciem melioris intellectûs gratiâ in superiori
Iconismo Z delineatam dedi) rudens iste *g y h* adeò distenditur
& attrahitur, ut necessariò inde utraq; Machinæ extremitas se
se attollere & erigere cogatur: id quod isto Instrumento, seu
Ergatâ perquàm lenissimè, nullo ferè labore, unâ manu peragitur,
ut sic mirum in modum exactissimè Machinam totam ad rectitudinem, si quid ei deest, reducere valeas, etiam sublatâ jam
Machinâ, ac in altum provectâ, omni tempore, quando volueris,
ac correctione & reductione aliquâ opus habere videris; quia
Ergata ad partem Tubi inferiorem residet, ubi nunquam non
illam arripere, manubriumque ejus circumducere Tibi est
promptum.

Eâ nunc ratione, atque hâc Ergatâ, unico hoc rudente à *g* per
k ad *h* ducto, integrum nobis est, totam hanc amplissimam centum quadraginta, imò centum quinquaginta ped. longam, maxi-
mæque molis Machinam quà ejus extremitates abundè angere &
constringere. Quòd si verò hæc Machina ad quinquaginta pedes
adhuc productior esset, & ad ducentos pedes excurreret (qualem
construere, adjuvante Divino Numine, me pariter posse utique
confido) rudens directarius *g k h* solus haud sufficeret, sed multò

longiori opus foret; quem itidem per quatuor trochleas, circa
extremitates Machinæ apponendas, prout circa intermedium rudentem primarium fecimus, nec non per alterum orbiculum inferiorem statuminis *F* ducerem, atque inter utramque rudentis
extremitatem Ergatam vel cochleam illam perpetuam annecterem,
quo adminiculo longè adhuc meliùs longitudini & ponderositati
Machinæ succurri posset : quin-etiam, si hæc nondum susti-
nendo oneri satis essent valida, intermedium rudentem gestata-
rium, per quatuor trochleas trajectum, per sex trochleas, atque
tertium orbiculum statuminis *F* perductare posses; quibus
omnibus adjumenta haud levia huic vastissimæ Machinæ impor-
tarentur : in summâ, ut videas, multiplici viâ talem molem me
exstruere ac regere posse. Præsens autem hæc nostra centum
quinquaginta ped. nec aliâ ratione, nec longioribus funibus opus
habet, siquidem iis, ut commonstratum est, optimè firmari potest
& sustentari.

Atque hæc breviter, quoad fieri potuit, de Machinæ hujus dis-
tensione & elevatione secundùm longitudinis suæ ductum dicta
sunto. At, inquies, hisce nondum expedita sunt omnia, nec
metam attingimus. Quippe cùm hæc constructio tantummodò
ex binis asseribus explanatis sit composita, licet quælibet hujus
Machinamenti pars quarta per se reverâ nulli inflexioni laterali
facilè cedat ; nihilo tamen minùs, combinatis & conjunctis ope
anginarum illis quatuor partibus, facilimè ob gracilitatem Machina se se notabili parte sine dubio incurvabit & obliquabit.
Utiq, largior, longitudinem illam nimiam, atque summam cor-
poris gracilitatem inducere aliquam incurvationem lateralem,
quando nempe in altum attrahitur ac commovetur; in primis,
si totam Machinam unâ extremitate regas. Id quod autem
necessariò, fateor, si quicquam accurati Tubo isto peragere sa-
tagas, summoperè evitari debet. Sed quomodo, inquies, huic
incommodo præveniendum? quàm facilimè, inquam, & quidem
leviori & planiori aliquo adjutorio. Nam, cùm maxima Machinæ infirmitas existat, ubi partium extremitates anginis
illis, cuneisq ; coadunatæ & constrictæ sunt, resticulam benè
distentam ad *m* alligatam pariter ad *o* firmiter annexi ; rursùs
continuando à *p* ad *r*, atque ab *s* ad *u* usque, ita ut asserum con-

junctio, & angina constringentes in medio resticulæ constitu-
antur, quo loco ligillum sive tigillum dentatum, quatuor ferè
ped. longum, sed subtilissimum abiegnum, vix uno & dimidio
pollice latum, crassum verò uno poll. ad basin Machinæ, sive
explanatum asserem applicui : prout ex præcedente Iconismo
A A ad *n q t* liquidum est. Deinceps, constitutis his ligillis
ligneis denticulatis instar sufflaminum ab utroque latere, tam
antico, quàm postico (pariter siquidem ibidem resticula dicto modo
annexa est) tum, quà parte videlicet Machina se se obstectit, resticulam à parte oppositâ, donec sufficiat, attraho & intendo, eamque
in crenam ligillorum convenientem, ne retrocat vel retrocedat,
intensioni immitto (planè ac si arcum intendas) & quidem
ab omni parte, ubi incurvatio se se offert, id faciendum,
usque dum deprehendas, Machinam ad lineam rectam omninò
esse reductam. Id quod facilè cognoscitur, etiam absque illis
ligillis visoriis, quæ loco Pinnacidiorum Tubis aliàs nostris, ut
Cap. superiori p. 390 & 393 dictum est, inserviunt, si duntaxat
juxta ductum longitudinis Machinæ ad quadratorū asserculorum
perforatorum cuspides collineas, protinùs omnis curvatura se
deteget. Quòd si verò animadvertis, intenso funiculo, exempli
gratiâ *n*, ab uno latere in excessu esse peccatum ; confestim
oppositam resticulam *rv* intende, & excessum recorrige, & sic
consequenter in reliquis curvaturis idem facito ; quò sic possis
per universa foramina ab initio scilicet ad finem usque Machinæ
liberrimè transpicere. Quo artificio non solùm Machinam ad
omnimodam rectitudinem rediges, sed etiam eam ipsam quà
partes circa combinationes & juncturas ita corroborare, ut appre-
hensâ unâ extremitate, ubi lentes oculares sunt repositæ, nullo
negotio alteram ab oculo aversam remotissimam ex voto commovere non nequeas, manente semper Machinâ planè incurvatâ
& illæsâ ; sed necesse est, ut tranquillâ aurâ hancce erigas,
tractesque Machinam ; ventus enim alioquin ob maximam pro-
ceritatem illam mirè vexat, jugiterque reciprocat.

Vides igitur, amice Lector, quomodo nonnunquam re aliquâ
leviusculâ, nullius ferè momenti, magno & arduo negotio
suppetiæ ferantur, si nimirùm debitè & cum ratione suo
tempore adhibeantur. Exemplum præstò est, quòd nimirùm

gracili resticulâ ad ligillum adeò subtilissimum & fragile serratum uno digito intensâ, tantam immensam molem leviusculâ
scilicet potentiâ regere & dirigere queamus, tantoperè, ut,
resticulâ in una aut alterâ productiori crenâ insertâ, illicò totam
Machinam notabiliter immutatam sentias. Totâ itàque
Machinâ ad amussim rectificatâ, tum quòque, an etiam in
elevatione & attractione ullâ aliquâ ratione à perfectione suâ
discedat, vel se se distorqueat benè exploratâ, binas illas cistas,
ex leviusculis & subtilissimis assamentis confectas, utpote *H* &
M, illam, cui tubulus diductilis duabus scilicet cicutis *z* in-
strictus est, hanc verò *M*, cui lens objectiva inserta est, ad
utramque Machinæ extremitatem tenaculis & subscudibus ligneis
adstringo, ne vel ullo modo nutent, nedum decidant; sed
oportet, ut omninò cognitum & exploratum in antecessum
habeas, quo loco superior cista *M* cum suâ lente convexâ, respectu lentium distantiæ, affigi debeat : quandoquidem Tubo
elevato tum primùm in distantiam inquirere, ac toties tantam
Machinam demittere, rursùsque attollere, nimis foret operæ.
Præstat ergò, ut jacente Machina veram diductionem ad ter-
restria objecta explores, atque inquiras.

His insuper omnibus peractis & observatis, tota Machina ad
elevandum, & phænomena exponendum stat in procinctu. Cùm
verò Machina hæc ingens sit moles, tum in longum jugiter excurrat, spatiumque, ubi conjungi, instrui & attolli debet, re-
quirat amplissimum; idcircò hanc ipsam in Speculâ nostrâ
ædium mearum haud tractare possum ; inprimis ob crassissimam
& procerissimam arborem, singulis vicibus erigendam. Coactus
igitur alium locum, atque receptaculum in prædiolo quodam
suburbano, haud procul ab Urbe nostrâ, in egregiâ planitie,
ubi undique liberrimus patet prospectus, elegi ; præprimis, quia
ibidem tam Machinam ipsam, quàm totum ejus apparatum
conservandi optima datur commoditas.

Hoc loco sub liberrimo Dio ingentem, crassissimam, pro-
cerissimamq ; arborem, nonaginta penè pedes longam primùm
in terram altè insertam, ac super decussatam basin, ex trabibus
robustissimis compactam, quatuor suis retinaculis renitentibus
impositam erexi & constitui ; ac ita quidem, ne ullo modo vi

procellarum commoveri, vel loco cedere possit, sed immota
semper consisteret. Per hanc arborem sursum versùs ordine
foramina ad climacteres in serendas sunt perterebrata, quò
fabro alicui lignario, vel alteri cuidam hujus rei assueto
ascendere, trochleamǫ; illam superiorem, binis orbiculis instruc-
tam, ad arboris cuspidem adstringere detur, atǫue postmodùm
funem ductarium illum longissimum, ac bene crassum per quatuor
orbiculos binarum trochlearum *x* & *y*, quò facilimè ac promptissimè moles moveatur, eò commodiùs trajicere liceat. Nam quò
per plures orbiculos funis ductarius transit, eò pondera minori
potentiâ, quanquam paullò tardiùs, moventur ac attolluntur.
His itaǫue instructis, necesse nunc est, ut trochlea *y* ad alteram
illam duplicem trochleam *c*, per quam aliàs rudens gestatarius
incedit, alligetur; sed hâc ratione id fieri oportet, ut statumen *k*
suo fune directario *g k b* inter cochleas dictas *y* & *c*, arboremǫue
remaneat : eam ob causam, ut statumen *F* à nullo latere declinet;
atǫue sic tota Machina à suo situ arrectiori vel minimùm
deviet. Id quod eò magis quoǫ; prohibetur, si maxima pars
trochlearum *a b c d* ad latus adversum respectu nostri, arborem
versùs, singulari tamen moderamine, alligetur; atǫue tum
statumen superducto suo & extenso fune directario quadruplici
rudenti gestatario, ne ullo modo unà cum Machinâ perverti
possit, securè incumbit.

Prætereà, infrà ad pedem arboris succulam, sive jugum
ligneum cylindricæ figuræ, quod duobus fulcris arrectariis transversim impositum, & suis vertitur cardinibus, decussatis, vecti-
bus adornatum locavi; cujus beneficio tota illa vastissima
Machina, duorum solummodò virorum auxilio, eousǫue, quò
phænomena exposcunt, promptè utiǫue elevatur. Postmodùm
in altum deductâ, tubuloǫue illo minori, duabus cicutis con-
structo, atǫue duabus plerunǫue lentibus ocularibus armato, ad
cistam *II* cochleâ suâ affixo, atǫue ligamentis probè, uti con-
suevimus, adstricto, eam ipsam Mensam nostram Telescopicam
versatilem, superiori Capite jam descriptam, Machinæ adhibeo,
eamǫue inter bina illa Mensæ illius statumina pari modo, uti
diximus Cap. XX, immitto, binasǫ; trochleas, funiculo ductario,
suoǫue ponduneculo, sive sacomate armatas, debito loco

appendo, non neglectis reliquis insuper adhibendis, promptum Tibi erit, etiam omnium maximum Tubum, utpote hunc centum quinquaginta ped. æque leniter, faciliter, ac velociter regere, ac ad quodvis objectum minimum non secùs, ac si Tubum viginti pedes longum, vel minorem haud majori sanè labore dirigere. Siquidem sacoma illud, & cochlea illa perpetua eandem operationem in maximis, quam in minoribus præstant. Tota enim Machina in suo centro gravitatis sustentatur, nisi quòd inferior Machinæ pars alteram partem superiorem ex præparato præponderet; quò sacoma, sive appensum pondusculum quinque vel sex librarum (quod tamen in aliis atque aliis Tubis variatur) in omni altitudine Machinam lenissimè conservare valeat.

De Mensâ autem hâc nostrâ Telescopicâ, ejusque usu quoniam jam præcedenti Capite abundè affatim diximus, pluribus hic supersedeo, sed Lectorē eò remitto. Cæterùm verò fortè adjicies: exhibuisti quidem Machinam duabus cistis, ad utramque extremitatem pro lentibus inserendis affixis; sed quomodò, quæso, tota illa Machina, ejusque partes intermediæ sunt cooperiendæ? præstabitne dicta Machina suum officium, cùm tota undique sit aperta? nonne lumen ad lentes, oculosque Observatoris pertransiet? refertne Machina hæc Tubum sive canalem undiq; occlusum? Respondeo, Machinam quidem non esse reverà Tubum omninò coopertum; nihilo tamen minùs prorsùs ejus fungitur officio, planè ac si reapsè undique esset tecta. Nam, cùm asserculi illi quadrati perforati non nisi tribus, vel quatuor circiter pedibus ab invicem tantummodò distent, omne omninò lumen prohibent à lentibus, oculoque Observatoris; ita ut dum oculum ad *H*, prius scilicet foramen, admoves, nonnisi obscurissimum, niggerrimumque Tubum sive canalem, atque in extremitate unicum solùm rotundissimum foramen deprehendis: asseres quippe isti omnes perforati ab illo antico latere, Observatorē versus, penitùs sunt denigrati, ut nil nisi Tubum, atro colore obductum, conspicias. Si negotium id exegisset, verum rotundum Tubum ex hâc Machinâ conficere mihi fuisset in expedito, inserendo scilicet tubos levissimos papyraceos, tres pedes longos, ab uno foramine ad alterum, &

ab unâ extremitate ad alteram; prout in hâc nostrâ Machinâ
circa ejus initium, incipiendo à cistâ illâ quadratâ *H* usèque ad
quartum foramen, ut ad *z z z* videre est: utièj; sic verum
Tubum undicque coopertum haberes. Adhæc aliâ etiam viâ
eundem effectum produxissem; si subtilissimo & levissimo
nigro linteo, vel velamento serico totam Machinam contexissem
ac velassem. Verùm his omnibus, ut ut facilè fieri potest, haud
opus tamen habemus; dummodò ad initio, incipiendo videlicet
à cistâ *II*, tria vel quatuor intervalla ad decem circiter pedes, ne
oculus Observatoris illicò ab initio divagetur, vel lumen extrin-
secum ad oculum pertranseat, tubulis claudantur, cæteris, ut
modò dicebam, optimè carere possumus, ut reapsè, si eadem
tentaveris, experieris. Ultimò de foraminibus hoc admonen-
dum adhuc habeo; quòd, cùm Machina sit maximæ proceritatis,
studiò foramina facta sint inæqualia; nimirùm præcedentia
paullò minora, subsequentia verò successivè aliquantò majora:
quò transpiciendo æqualia omninò appareant; aliàs profectò, si
fuissent prorsùs æqualia, illa maximè ab oculo dissita longè
minora apparuissent. Ideòque priora nonnisi octo poll.; subsequentia verò novem, decem ad undecim usèque poll. confeci.
Quanquam parùm admodùm interfuisset, an æqualia planè ex-
titissent, dummodò nihilominùs in rectâ lineâ Machina conser-
varetur, ut optimè transpicere potuissemus; attamen præstat,
ut antecedentia sint aliquantò cæteris vicinioribus ampliora:
eam quidem ob caussam, ne protinùs, dum Machina se se
parumper incurvaret, prospectus impediretur. Accedit, si præcedentia sint ampliora, parùm referat, an Machina circa medietatem se se paullulùm inflectat.

Postremò, quemadmodùm omninò laboriosum est, nec sine
aliquo sumptu, temporisèj; dispendio fieri potest, ob plurimorum
virorum operam, quâ carere haud possumus, vastissimam hancce
Machinam conjungere, constringere, instruere, attollere, commovere ac dirigere; sic quoèque eandem demittere, restringere,
disjungere, apparatum adimere, omnia suo loco deducere & conservare, labor est haud vulgaris: verùm in hocce negotio fieri
aliter non potest, nec cuipiam Siderum Cultori hoc debet esse
molestum & tædiosum. Nam cùm arbor in patenti campo con-

sistat, Machina cum trochleis, funibus, totàq; supellectili
ibidem haud remanere potest, sed peropus est, ut omnia &
singula relaxentur, removeantur, ac destinato loco ordine
reponantur; secùs profectò ea omnia, quæ haud leviusculis
sumptibus, laboreque comparata & constructa fuére, facilè ab
injuriosâ temporis tempestate consumerentur, pessumque irent.

Quomodo autem hæc molestia evitari possit, jam Capite
superiori meminimus; si nimirùm locus aliquis commodus, ac
his Machinis & Observationibus aptus instrueretur. Verùm id
ipsum ad effectum deducere, non est privati alicujus hominis,
necessum est, ut sit Princeps aliquis, magnus Rerum Cœles-
tium Mecœnas, qui sumptus liberalissimè suppeditaret, totumq;
negotium promoveret; utiq; sic confido, Speculam atque Observatorium fundari atque adornari tunc posse, quò nunquam non
quocunq; tempore, quotiescunque libeat, ad paratissima plenè
obarmata Telescopia, sive sint viginti, quadraginta, sexaginta,
centum, imò centum quinquaginta pedes longa, accedere liceat :
ut nihil quicquam ampliùs opus habeas (cùm funes ductarii suo
loco semper explicati pendeant) quàm illud, quodcunq; velis,
imò duo, tria amplissima simul attollere, & ad Astra exponere ;
protinùs vicissim, quando volueris, quodcunque cum toto
apparatu, nihil quicquam ei adimendo vel immutando, debito suo
loco reponere, ubi tutò singula quiescere possunt, & minimè in-
temperiei aëris obnoxia, nec alterum alteri, nec Observationibus
ullo modo sunt impedimento : cæteras commoditates ut modò
præteream, de quibus autem omnibus proximo Capite XXII
plenè acturi sumus, atque Designatione evidenti, rem ita succe-
dere posse, commonstrabimus.

CAPUT XXII.

De Peculiari Observatorio, maximis Telescopiis convenientissimo.

Ejusmodi Speculam, pro omnis generis Telescopiis longioribus Cœlo admovendis & explicandis aptissimam modó detegendá ratione commodè exstrui, & necessario apparatu instrui posse, penitùs persuadeor. Verùm, prout jã suprà tetigimus, haud res est alicujus hominis privati, sed magni Principis, cui nec commoditas loci, nec opes, multò minùs ardor, adeò sublime negotium Rei Sidereæ bono promovendi, deessent. Nam, crede, operosum & sumptuosum initiò foret opus; attamen semel adornato, maximas incommoditates in hocce Observationis negotio evitare promptum esset; sic ut semper ad paratissima accederemus, atʠe nunquam tanto temporis dispendio, tantàʠue molestiâ procerissima ac ponderosissima illa Telescopia priùs componere, atʠue instruere opus haberemus.

Exædificetur igitur ex mente nostrâ in loco quodam undiʠue patentissimo ac liberrimo Turris quædam rotunda, haud tamen fastigiata, sed ab omni parte ab imo ad verticum usʠ; æʠue crassa, sive ex trabibus compactilibus, sive ex laterculis, cœmentitio nempe opere (sicuti cuiʠue videbitur) altitudine 100, vel 120 pedum, si nimirùm Telescopiis 150 pedum qouʠue inservire debeat; secùs altitudo dictæ Turris utiʠue minor esse potest. Latitudo ejus sit ad cujusvis placitum, pro commoditate negotiorum atʠue requisitâ proportione; quanquam, si nulla alia habenda est ratio, poterit esse quàm arctissima. Quippe si sola Telescopia respicienda sunt, sufficit, si diameter ejus ad 12, vel summùm 15 ped. excurrat; non tantùm enim spatium interius pro iis omnibus, ex nostrâ sententiâ ibidem expediendis, satis tum capax esset, sed quoʠue ad diversas concamerationes & contignationes exstruendas.

Quò autem institutum meum eò pleniùs, clariusʠ; cognoscas, lubenter illud in Delineatione appositâ B в Tibi distinctè sub adspectum ponere volui; ex quá nemini non, Rebus vel leviter

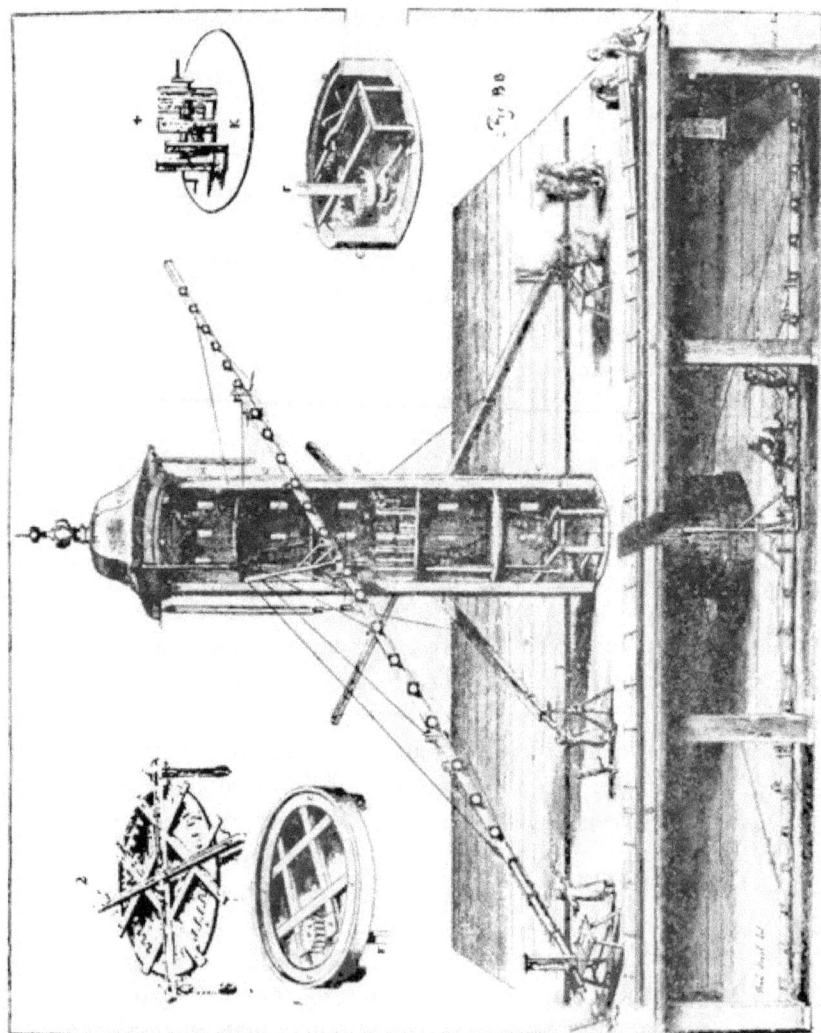

Fig. 98

Mechanicis imbuto, totum negotium abundè illicò intelligere promptum erit. Nimirùm exædificarem, ut modò dicebam, in campo, vel areâ quâdā liberrimâ ejus generis Turrim, ut vides A, eam u e circumcircà in altum aliquantò assurgente Tabulato quadrangulari, certis Fulcimentis ligneis incubante, instar alicujus Procestrii sive Theatri spatiosissimi instruerem; cujus latera, ut longitudinem Telescopiorum videlicet adæquent, minimùm sint 150 pedum; altitudo verò hujus Proscenii esto 15 vel 18 summùm pedes, quò erecta Telescopiorum statumina intermedia, super quæ rudens adstrictarius aliàs ducitur, ibidem consistere possit. Columnis vel statuminibus robustissimæ trabes eatenus imponendæ sunt, quò crassiores asseres iis imponi, totumq; Procestrium contegi commodè non nequeat. In hocce autem Tabulato propè ipsam Turrim exscindatur apertura quædam, ut pateat interstitium, quò Telescopium, in pavimento inferiori quiescens, & cum omni suo apparatu instructum, attrahendo liberè pertransire, ac posteà rursùs tegumentis ligneis, ne pluvia, vel quisquam Observatorum & Spectatorum in concamerationem istam, in tenebris obambulans, fortè incidat, operiri possit.

In ipsâ verò Turri tot contignationes, quot volueris, ac opus habes, Tibi adornare licet. Inferius conclave sit promptuarium supellectilis Telescopicæ, reliqua superiora ad res diversas, ut mox pleniùs dicetur, possunt reservari. Oportet autem ita inprimis dicta Turris sit constructa, ut totum ejus tegmen superius, unà cum quatuor illis transversis trabibus gestatariis, ad quas alioquin trochleæ suspenduntur, se se queat convertere. Nam quia Telescopia, licet semel sint ad phænomena directa, haud uno eodemque situ loco, nec sub unâ eâdemque elevatione detineri & conservari possunt, sed opus est, ut continuò secundùm diversum objecti motum, situmq; obvertantur ac dirigantur; hincque necessum est, ut tegmen etiam hujus Turris adeò sit concinnatum, ut quàm facilimè quaquaversum dirigi, atque haud multo labore unius aut alterius famuli auxilio circumgyrari queat.

Ut autem eò clariùs omnia percipias, haud gravatus sum per diversa Schemata rem omnem Tibi ob oculos ponere. Primò

igitur scias, me vertici parietis hujus Turris, ubi culmen initium ducit, orbem robustissimum ligneum, parte excavatum
peculiari scilicet striâ, sive projecturâ elaboratum, ut ad Num. 1
videre est, imposuisse, ad quem prætereà orbē quatuor transtra
orthogonaliter affixa sunt : partim ut orbis hicce ligneus eò
esset firmior, partim ut commodè axiculus striatus ligneus *a*,
beneficio alicujus asseris *b*, ad orbis peripheriam adaptari conce-
deretur. Ad *c c* verò orbem eo fine excavavi, ut alius orbis
ligneus, ab interiori parte dentatus, ejusdem magnitudinis, sicuti
ad Num. 2 patet, intromitti, atque in eadem striâ *c c* circum-
duci posset. Quò autem hicce orbis eò redderetur robustior
(quippe non sòlum transversas illas quatuor trabes gestatarias
f f f f, ad quas trochlea ac funes unà cum Telescopiis tantæ
molis annectuntur, sed & totum culmen sustinet) quatuor insuper transtris *e e* decussatim affixis (quibus alioqui tigilla
culminis innituntur) orbem istum *d* corroboravi ; ne non totum
tegmen proclivius, unà cum trabibus gestatariis, Telescopiis,
totoque apparatu, & quidem unius aut alterius hominis auxilio
in gyrum ageretur.

Hocce negotium verò ad eò promptiùs promovendum, axiculum striatum *a* ita adornavi, ut aptè dentes majoris rotæ *d*
arripiat, atque sic horsum vorsum rotam ipsam circumducat.
Axiculum autem striatum *a* haud opus est Ergatâ quâdam,
sive Machinâ attrectoria ex superiori aliquâ contignatione H
vel G convertere, sed perinde est, quovis loco id fiat ; hic in
appositâ Figurâ B n iste axiculus *a* ex conclavi C, ubi eo fine
succulam trabeculis transversim fulcris arrectariis adaptatis im-
posui, aptè regitur ; cui porrò succulæ sive jugo, quod in suis
vertitur cheloniis, alius axiculus striatus *b* inditus est, conver-
tendus nempe à rotâ majori coronatâ *g*, per cujus centrum arbor
procerissima F instar fusariæ unâ cuspide, rostro scilicet suo,
in cheloniâ *k* convolubilis trajecta est ; altera verò illius arrèctæ
arboris, per omnes contignationes ascendentis extremitas superiùs
axiculum striatum *a* gerit ; quò sic major illa rota superior *d*,
unà cum trabibus illis portatariis *f f*, reliquisque illis quatuor
e e e e transversariis, quibus canterii culminis insistunt, commoveri, ac in gyrum duci possit.   Id quod nunc manubrio

quodam *m*, sive vecte haud multo negotio à quolibet peragitur, ut levissimè ac ocyssimè Telescopia etiam vel maxima, ac ponderosissima omni occasione, & ad omnem Cœli plagam dirigere ac convertere nonnequeamus. Quòd si arborem breviorem huic negotio adhibere mavis, Ergatam ex conclavi C ad superiorem aliquam contignationem, sive concamerationem G vel H transferre conducit.

Ad Telescopia ipsa verò sustollenda, & in sublime attrahenda, peculiarem Machinam attractoriam, uti vides in contignatione E, vel ut clariùs conspicitur ad Num. 4, ex quatuor peculiaribus succulis condidi ; quò sic etiam quatuor Telescopia simul in altum arrigere liceat. Nam cùm à quatuor diversis trabibus trochleæ cum suis funibus dependeant, necesse quoque est, ut tot Machinæ attractoriæ etiam ad quemlibet Tubum elevandum & dirigendum adsint ; funis ductarius autem per trochleam *f*, & rursùs per *n* ductus est, ita ut omne negotium sub tecto in ipsâ Turri peragi possit, exceptis ipsis Observationibus, quæ ab Observatore ipso sub dio in patentissimo illo Theatro expediuntur. Atque hâc ratione nunc sub illo spatiosissimo Tabulato, ab omni parte clauso, quævis Telescopia omni apparatu instructa, omnemque etiam supellectilem confervare commodè datur ; ac rursùs datâ occasione, quoties lubet, quodvis, cujuscunque etiam sit longitudinis, protinùs demisso scilicet ex trabibus illis gestatariis *f* fune ductario, atque Telescopio debitè iis alligato, ex cryptâ vel crypto portico extrahere, atque ad Observationes pro lubitu haud multo labore explicare : adeò, ut non solùm unicum Telescopium, sed simul quatuor utique ad Astra dirigere possis, imò, si opus foret, plura, dummodò ad rotam *d* totidem trabes portatarias affigas & accommodes.

Quo pacto verò quodvis Telescopium funibus constringatur, attrahatur, tum ope Mensulæ directoriæ ex voto regatur, haud attinet hic pluribus recensere ; siquidem jam suprà suo loco satis de iis omnibus fusè disseruimus, ut nihil ampliùs restet, nisi quòd admoneamus, quòd peractis Observationibus nihil, nisi Telescopia per aperturam, sive ostium B & I in concamerationem demittere, funesque ex superiori contignatione adimere, deniq̃; operculis eò destinatis Tabulati aperturas contegere opus habeas, atque tum omnia & singula ita condidisti ac custodivisti, ut non solùm ab omni corruptione, aërisque injuriâ sint salva & sarta tecta conservata, sed etiam quovis tempore omnia paratissima rursùs habeas, quando ad Observationes iterum est redeundum. Quâ equidem viâ omnem illam gravissimam molestiam evitare possumus, ut minimè necessum habeamus, tantas immensas Machinas primùm ex omnibus latebris conquirere, ex tot ac tot partibus componere ac combinare, tot funibus, trochleis ac Ergatis construere, constringere & adornare; nec non paullò pòst, expeditis Observationibus, unâ aut alterâ horâ elapsâ vicissim omnia & singula relaxare, disjungere, suoque loco, maximo Observatorum labore, temporisq̃; in primis dispendio, transportare. Quibus de caussis sæpiùs, ob molestissima scilicet & sumptuosissima hæcce negotia, plurimæ egregiæ Observationes ab Observatoribus deseruntur & negliguntur, quæ alioquin sanè, si ejusmodi commodissima daretur Specula, maximâ cum voluptate ac Astronomiæ commodo, neutiquam interirent, vel inobservatæ elaberentur.

Verùm enimverò, quis, fortè inquies, privatorum huic operi construendo par erit? Nam, ut mihi ipsi ultrò fatendum est, illa ipsa Turris non nisi à Principibus exstruitur; attamen ut videas, etiam alicui privato homini mediocris fortunæ, ut mihi & Tibi, id non omninò esse impossibile, planiorem adhuc rationem hoc loco detegam, quam etiam longè levioribus sumptibus nobis comparare possumus, eâdem ipsâ, ut diximus, commoditate ac utilitate, pari ferè modo, ac si talem splendidissimam ac sumptuosissimam Turrim, cum suo patenti Theatro vel Proscenio exædificasses. Hoc nimirùm modo; si ad pedem arboris Cap. XXI pag. 415 descriptæ, atque in arcâ quâdam patenti erectæ, ubi Tibi commodum videbitur, concamerationem quandam, ex asseribus & statuminibus undique contectam, exstrueres, tantæ scilicet longitudinis & latitudinis, ut Telescopiū aliquod 150 ped., si eo gaudeas, in eo conservari & contegi posset. Hocce autem conclavium necessum est ut sit ab omni parte clausum, & à superiori parte, ut diximus, asseribus benè tectum & consertum, nisi quòd certas suas excisiones sive foriculas à superiori

parte habeat, quò Telescopiis, quemadmodùm in superiori
Theatro ad B & I conspicitur, liberrimus detur transitus. Quod
conclave sufficeret, meâ opinione, pro conservandis Telescopiis,
omniàque eorum apparatu necessario, ut vix quicquam hâc ratione
desiderandum habeas, nisi quòd in quâvis Observatione funem
ductarium cum suis trochleis ab arbore istâ arrectâ dissolvere,
ac deportare opus sit : ut negari haud possit, nos plurimas difficultates & molestias hoc modo evitare posse.

Verùm Observatorium hocce longè adhuc aptius, com-
modiusque redditur, si ad arborem tantam fossam ducas, quò
Cryptam subterraneam, sive Cryptoporticum, quæ longitudini,
latitudini & profunditati ad condenda & asservanda Telescopia
sufficeret, condere posses. Cujus autem Cryptæ subterraneæ
latera aut ligno munire, aut laterculis cœmentare oportet ;
superior verò pars asseribus arctè ad pavimentum tegatur, ita
tamen, ut foriculæ remaneant, per quas Telescopia extrahantur,
ac rursùs immittantur. Quâ viâ, quicquid hâc in parte
desiderare, vel ab illâ superiori sumptuosâ Turri exspectare
posses, ut puto, in promptu haberes. Integrum enim Tibi
foret, totum apparatum in Cryptoportico illâ subterraneâ, longè
minoribus sumptibus constructâ, condere ; ut nihil omninò
ampliùs superesset, quàm foriculas aperire, funes ad arborem
alligare, & tum ad paratissima & instructissima varii generis
Tubospicia accedere. Intereà tamen, si hæc Observatoria modò
à me detecta Tibi, Benigne Lector, fortè nondum arrideant, alia,
sis, in commodum Divæ nostræ Uraniæ adinvenire, atque nobis
quòque detegere poteris : facies rem, crede, Astrorum Metatoribus
omninò longè gratissimam.

TRANSLATION.

CHAPTER XVIII.

Of a Peculiar Helioscopic Instrument for the accurate determination of Eclipses and Solar Spots.

BY the assistance of this Helioscopic instrument, some time since invented by me, and pretty fully described at page 98 of my " Selenography," I imagine that all will readily allow, as experience too has shown, that solar spots and eclipses can be more precisely observed and delineated; and as the entire apparatus of this helioscopic instrument (for instance, the revolving or turning sphere in the little aperture, with the bench, table, and all their appurtenances) has been quite clearly explained in the same place, I would say nothing more of them here, but refer the reader to that page. Still, as the aforesaid instrument illustrated in my " Selenography " is, as it were, the foundation of the present new one which has to be described, I consider it worth while to annex at least a figure of the same, so that not only will the description be in itself much easier for me, but the entire construction of the new instrument will be much easier for you to understand. You have, then, kind reader, in the annexed plate an accurate illustration of the former Helioscopic instrument. I have no doubt that astronomers will perfectly understand, without much difficulty, in what way solar spots and eclipses should be observed, and if some points happen to be not sufficiently clear, it will be easy to clear them up from the prolegomena of my " Selenography." The plate, indeed, very sufficiently exhibits many parts which would have otherwise to be described, parts which I still retain in the new. Notably, amongst others, I still retain the same turning sphere, the bench with its two vertical supports spirally grooved, and two worms for them to work in, by means of which the table so supported can be raised or depressed for any altitude of the sun, and also the telescope fitted with a tube, and brought to a right angle with the table by means of a square bar, so that the sun's rays may fall at right angles to the table, and its disc be portrayed on the paper; so that I have not made any improvement up to the present time on the method of taking observations except in one single particular, i.e., in the plan of the table b, on which the sun's reflection is received, and upon which is the paper, with the circle or disc s v v w z on it. This instrument is certainly, as I have proved, most admirable for intercepting and marking down solar spots and eclipses, were it not that the sun, by reason of its very swift movement, especially its diurnal movement, by which it is not only constantly ascending and descending, but also moves from west to east, (to speak after the manner of the peripatetics), and is in such continual motion that it never for an instant stops at rest on the table, or rather on the disc, and consequently when any accurate observation has to be made, you are obliged to keep the table in constant motion, and to keep turning the revolving sphere first this way and then that; at one time to elevate it more and more, at another to depress it. When thus engaged, unless you have some one with you very well versed in the use of instruments, and quick-fingered, you are but measuring sand. Worse still, if you are alone, and have to keep constantly moving the instrument, and at the same time marking down the phenomena. I, then, clearly seeing all these inconveniences, so early as the year 1661 applied to the eminent Bulliard, in those days one of my most valued friends, when he was with me on the occasion of an eclipse of the sun, as to how I could best obviate these difficulties, and cause the sun's reflection to remain steady on the object-table, and in the circular disc described on it, or, what comes to the same thing, how the sun's image might be kept constantly without motion on the circular disc, in order to mark down spots and the different phases of eclipses in the most accurate manner, and without any vibratory movement. The illustrious Bulliard strongly advised me at that time to carry my project into execution as quickly as possible

D

for the benefit of all observers. Several men well acquainted
with mathematical subjects in general were of opinion that the
project was not only one of great difficulty, but even an almost
impossible one to carry into practice. To me, however, it did not
seem so absurd, since I considered that as in the manipulation
of the *pinnacidia* of quadrants, I had been able to keep the
sun for certain from ever quitting the disc, so by the Divine
assistance I should undoubtedly carry this project also to a
successful issue; and so it turned out, for the matter shortly
after succeeded to my heart's content. The whole secret of the
device lies in the table connected with the revolving sphere, or,
if you like it, with the telescope, for the first table belonging to
my former little Helioscopic instrument, described in my
"Selenography," a table which rests merely on the bench, can
be moved to suit only the same constant longitude of the
sun, unless you raise or depress the board by means of the
screws on which it is fixed.

Now, to do this with a simultaneous movement of the table,
so quickly as to keep pace with the sun's motion, is scarcely
ever possible. On the contrary, the newly invented table
which I at present use, though the bench be kept always at the
same height and in the same position, can nevertheless, by the
arrangement of two very small screws, be readily and most
exactly directed according to the sun's motion, so that the sun's
image remains constantly within the disc, and does not quit it
even for a moment. The circular table spoken of (g) is made
of different parts, which I will describe, so far as shall be
possible, separately, in order to be better understood. Such
particulars, however, as I shall not be able so clearly to set forth
will become clear from the annexed diagram of the several
parts, in which they are distinctly portrayed, or if you prefer it,
you will easily understand them more by your own natural
ability, and by going through the arrangements thereof for
yourself. First of all, it was absolutely necessary that the table
(g), on which the paper, with its observatorial disc, has to be
fixed, should be able to be moved as smoothly and as quickly as
possible this way or that, in accordance with the sun's azimuthal

as well as vertical motion. Secondly, it was no less a matter of the first necessity that the same table, with its disc described on it, might with equal readiness and quickness be moved according to the sun's motion in right ascension; that is, in a circle. To accomplish this properly two very peculiar wooden wheels of different size are, in my judgment, requisite, so that the larger (g) may be joined to the smaller (c) by means of an axle (m) if the two motions, the rectilinear and the circular, are to be obtained. I took, therefore, two very thin boards of linden wood, about sixteen inches long, as you see at a, and glued them together at right angles. To the upper part of the vertically placed board a metal plate, with a long, endless screw (b) attached, passing through two catches, was fastened by some small screws. Then through the smaller wheel (c) I passed an iron spindle (m), fitting on to the back part of it a cogged axle (i), and so I passed the spindle (m) along with the entire wheel (c) through a hole (k) in the vertical board, and firmly screwed them to the back part of that board. When these parts are thus joined together, and the endless screw turned, since its threads work in the cogs of the axle (i), the smaller (c) wheel must turn round as the endless screw turns and twists in one direction or the other. Then in the front side of the smaller wheel (c) I fixed a peculiar little piece of mechanism (d), quite similar to that which (Chapter V. p. 115) I had formerly applied to the horizontal quadrant, or rather to the bar to which that quadrant was at that time attached, for the purpose of bringing it to equilibrium, or of adjusting and rectifying its level; but as this mechanism has been fully described in that place, I will not dwell longer on it. It consists, for the rest, of a single small screw (e), which moves the solid intermediate part of the mechanism up and down. To this intermediate part (d), to which the aforesaid screw (e) is attached, two little rectangular brackets (ff) are nailed on, in order that the greater circular table (g) may be supported by them, and strongly fastened to them by small screws; and then by combining the two wheels (c and g) it will be in the power of any one by turning round the screw (e) to move the greater wheel up and down for the

vertical motion, at his pleasure; by turning the other screw, the
endless screw (b) to make both wheels (c and g) revolve together,
and so smoothly and readily that scarce any motion can be more
exact or swifter. And this is the whole construction of the new
Helioscopic instrument, by means of which it will be in your
power, with scarce any trouble to yourself, to keep pace most
exactly with the whole motion of the sun as regards both longitude and latitude, so that by this method the sun's image is
forced to remain quite stationary within the observatorial disc,
as I said, and not to quit it even by a nail's breadth, as will be
presently more fully stated. It is further necessary that you
attach the entire little machine to a stick or square bar (h) by
inserting the bar through a square hole (a), so that the
aforesaid machine, being firmly wedged on, may remain
perfectly steady at right angles with every movement of
the tube, or telescope appertaining to the revolving sphere and
inserted in that tube. Then the entire apparatus, with
its two wheels attached, the lesser (c) and the greater (g), on
which is nailed the paper, with the observatorial disc described
on it, must be placed on the Helioscopic bench, as illustrated in
Plate W, and must be roughly directed to the sun by moving the
wheels this way or that by means of the screws, so that the
solar rays may almost fill the disc (n). Then, all being ready, I
bid my assistant mount the steps behind, and he, seizing with
his left hand the screw (e) which turns the table along with the
observatorial disc, according to the motion in right ascension,
with his right the other screw (b), which turns the smaller wheel
(c) along with the larger (g) in a circle, will be able, by turning
one screw or the other this way or that, with wonderful ease, so
nicely and so gently to neutralise the whole motion of the sun,
that its exact image will not in the slightest degree pass beyond
the limits of the observatorial disc, but be, as it were, immovable
upon it; so then the assistant to the principal conductor has
nothing more to do than to keep his attention constantly fixed
on the representation of the sun's disc and the observatorial
circle, that the rays may not at any point pass outside the circumference, but constantly remain perfectly enclosed therein.

It will be easy enough for any one in the world to do this, and the plate above makes it thoroughly intelligible; and in this manner it will be easy for any observer to mark down solar spots, and eclipses too, most readily and correctly, free from any trouble in the direction of the instruments, and from all hindrance by reason of the sun's motion, in the most agreeable manner, as if in sport, and certainly with far more exactness than has ever yet been considered possible, and as often as he pleases, to resume and correct an observation. Now that I have finished this sketch, I must further add that it is necessary to mark a vertical point on the edge of the upper part of the observatorial disc, in order to obtain the inclination of the sun by means of a movable string pendulum, hung upon a rack projecting from one of the brackets, as I have already fully explained in the proper place in my "Selenography." Lest, however, the entire Helioscopic instrument, with all its iron apparatus, should fall flat, or by its too great weight should bend or twist the bar (h), and with it the telescope itself inserted in the hollow sphere, it is requisite that the assistant, who, of course, manages the screws, should at the outset shore up the whole machinery with a prop (o) at the back, inserting the other end of the prop slantingly into the floor, and thus you will keep the whole apparatus securely in its true position so long as you please. For the rest, I should wish you to notice that though the wheel (g) is large enough—it is sufficient, indeed, if it have a diameter only of a foot—to require in this new Helioscopic instrument two levers (h and l), l the shorter and h the longer, as is very clearly shown in the plate, so that the point needs no elaborate explanation, the management does not from this prove at all more difficult, or even the observation, than with the old single lever, but all is performed with equal facility, for both the wheels as well as the revolving sphere obey the movements of the two screws, and turn as you please. It is not for me to expatiate here on the suitability and accuracy of this little piece of mechanism, especially for marking eclipses of the sun, but others who will some day use it for observations will no doubt make ample acknowledgment of its utility. I feel quite convinced that this method has been unknown as yet up to the present day (forgive the verbiage), much less practised; and therefore I hope it will not be so very unwelcome to astronomers, whose observations in such an exalted field can never be too exact. They should take care to examine every point aright, and to have the machinery made and put together according to the diagram and the description here given, or even according to the description already published in my "Selenography." I leave this subject, therefore, to prepare myself to treat of telescopes and optical tubes, since they afford, as will be abundantly clear from the following chapters, invaluable aid in astronomical discoveries and researches.

CHAPTER XXI.

Of my Great Telescope.

SINCE it is now, though late, accepted as a proved fact that nothing furthers the interests of science more than the use of accurate and perfect telescopes, by means of the numerous celestial observations they record, every astronomer and optician of the present day wishes nothing more heartily than to make them day by day more perfect and of greater length. Yet telescopes more than sixty or seventy feet long, though lenses for longer tubes may have been polished here and there, have scarcely anywhere been applied to the stars with any tolerable success, for the reason that makers have not up to the present time been able to invent and put together suitable machinery, and to fit up a proper tube to hold the lenses, such a one as would be readily obedient to every movement and direction, and give as little trouble as possible to the observer from its weight and bulk ; and the difficulties are such that most scientific men have almost despaired of any simplification of means to render the invention possible. Nevertheless, I did not by any means cast away all hope, for I felt perfectly confident that by the blessing of God a very large instrument of the kind, for holding lenses, might be constructed, a hundred or a hundred and fifty feet in length, and be turned to good purpose ; and that illustrious gentleman, Titus Livius Burattinus, an expert in all kinds of mechanical as well as optical studies, specially kindled my enthusiasm to undertake the task, not only promising lenses polished by his own hand, but even a short time afterwards generously producing them. Still, at first I confess, I plainly foresaw that there were very many difficulties in the way. For were I to set about the business according to the old plan, i.e., by constructing a square tube of four planks, and joining them into four lengths, one length would be but forty feet. These separate lengths might, indeed, be joined together by some more complete or more novel method than any before discovered, and,

by means of boxes and ropes, be formed into one straight piece,
if indeed, one were not afraid of extraordinary expenses, and the
want of machinery adequate to put together and direct the
instrument. In truth sixteen planks, each forty feet long, and
of sufficient thickness and breadth, together with three very
stout boxes, each at least fifteen feet long, protected by such a
quantity of iron, with the addition of so many ropes and pulleys
and other apparatus—what a task! what a toil it would be to
join all these together! to bring the instrument to a level! to
elevate it! to move it! and to keep it constantly in its proper
position and free from flexure! For these reasons, I, foreseeing
the extraordinary difficulty of the enterprise, tried to succeed by
an entirely new method; and this was to construct an instrument for lenses one hundred and fifty feet long, of only eight fir
planks, and these much narrower and finer than ordinary. The
instrument would then be lighter, and, therefore, much more
convenient and easy to put together and manage; notably, as the
intermediate boxes would be entirely dispensed with; and, which
is most important, it would be put together without using a
single farthing's worth of iron. To be brief, I studied the
construction of an instrument to serve the purpose of a round
tube, in parts; each part to consist of only two rather slender
planks. The idea appeared to some, by no means ill-versed in
such matters, absurd, nay, even paradoxical—to build up and
put together a tube by merely joining together two perfectly flat
planks. Since it is quite indisputable that two smoothly
planed planks could never be combined and joined so as to
represent a closed tube and exclude light at every point.
Nevertheless, by the help of God the work prospered, so that I
succeeded in the object desired, namely, in the ability to use the
instrument handily, as if it were a veritable tube, although, as
you can easily understand, it is a matter of some difficulty, not
only to put together this said instrument, but likewise to
manage and direct it. I built the instrument in four pieces,
each forty feet long, in order that I might obtain the required
length of about one hundred and fifty feet. First of all I fixed
a deal plank about ten or eleven inches broad and one and a

quarter thick, evenly planed and side uppermost, at right angles upon another smaller one, only eight or nine inches broad, and scarcely three quarters of an inch thick, applying it to the middle of the side of the smaller one, as to a base—the side being first planed perfectly even, as can be seen from the annexed diagram at $a\,B\,a$. And in order that these two planks might be more strongly joined I attached numerous wooden props and braces lengthwise (c), at intervals of three or four feet; applying one side of the props (c) to the base, i.e. the lower plank (b), the other side to the plank standing on it at right angles ; and to the outside of this plank ; clenching them also, and rivetting them most strongly with wooden plugs. Thus I gave them such stability and strength that these two planks, the lower one and the upper one, could not stir or twist away from one another, but must always keep united in that position. For, as I have already mentioned (Chapter VIII.) with reference to the quadrant, that the board $a\,B\,a$, standing on the other smoothly planed one ($b\,b$) in no wise allows the lower—the base—to curve, although the plank is in itself quite narrow and slender, so contrariwise, the underneath plank ($b\,b$), we are now speaking of, when it has once been throughout evenly attached to the upright, nowhere allows the upright to deviate from the perpendicular, so that these two planks, each forty feet long, when joined and fastened together in the manner described, are made to preserve a perfectly rectangular position, and can by no means be bent; nay, if very great force were employed you would sooner break the planks thus united than make them diverge. And this may easily be imagined. On the other side of the upright plank ($a\,a$), as may be seen at A and D, i.e., on the side opposite to the supports and braces (c), I placed square boards (d), also at right angles, about a foot square, having a large hole bored through them, and fastened them tightly, and rivetted them with wooden plugs to either plank, the underneath one (b), and the upright one (a), not only with the view of giving additional support and strength to the planks, but principally—not to mention other things at present—entirely to exclude all light from the outside, which might fall on the lenses or dazzle the eye of the observer. And so the four parts, *A B C D* being thus joined and fitted together, we must now find means to join and bind together the upright boards lengthwise, so as not to use any iron whatever, and yet that the work may possess the requisite strength, and be also quickly unfastened and taken to pieces without much trouble and loss of time. Now in order to ensure this, I have not been able to hit upon any better device than to make a special kind of wooden cramp, such as boxmakers use when they join and fasten boards together. Of such I have given illustrations in the preceding diagram, and in the sketch to the right hand. There is this difference only that the cramps of boxmakers are strongly joined together and rivetted at both ends, and whatever has to be bound and compressed by them has to be inserted in the aperture, or between the limbs; while my cramps, for keeping the instrument together, are closed and rivetted only at one end. They are closed at the other by a peculiar square wooden bolt with bevelled edges, inserted as a catch in the grooves at each side of the cramp. A wooden pin is then passed through to prevent the bolt slipping out, and to give it more resisting power, and also that it may be instantly taken out whenever this is requisite. Into a pair of these cramps (*f*) the ends of the two upright planks *D* and *C* are together firmly inserted at their top part, the bolts being first removed from the cramps, the size of which exactly corresponds to the thickness of the planks. A better plan is to pass the two cramps (*f*) through slits of sufficient size in the under-plank (*b*), and then to fasten them on the top. The two cramps are thus closed by the square bolts (*e*) so that they cannot come off or break away, and are secured by their pins. I afterwards insert into each cramp, and hammer down, two very stout wooden wedges of exactly the same breadth as the cramp, but with their edges in opposite directions, between the fastening (*e*) and the two united planks, and drive them with a rammer until they will give no more, and till both planks *D* and *C* are sufficiently bound and tightened together. This is illustrated at *E*, best from the figure above to the right hand : and the two portions of this

instrument are so compactly united that they cannot be dis-
jointed without taking out the wedges. In like manner the
remaining portions A and B are, in their turn, fastened together by four similar cramps and wedges. So that all four
parts form by means of three pairs of cramps one structure in a
perfectly straight line. You must now well observe that in the
centre of the instrument at E, a rest F, erected on a cross base
G must be placed on and set up at right angles to the instrument, and by clamps and wedges immovably fastened to it.
The aforesaid rest is twelve feet high, and well strengthened by
two props and braces, as is clear from the figure above. On the
top of the rest two wheels turning on an axle are fixed at i and
k, for what purpose shall be presently explained. All these dis-
positions, then, being made, and the different portions of the
instrument being firmly united, the rest also standing on the
top upon its cross base, the entire instrument is conveyed to
the mast, on which it has to be elevated, and being there extended forwards on the steps projecting horizontally from the
sides of the mast, after its level has been proved to be perfectly true throughout, by the removal of any curvatures, so
that an uninterrupted view can be obtained through the holes (d)
from one end to the other, ropes and cables of different kinds, as
well as many pulleys of different sorts are attached and firmly
fastened to the instrument, as is plain from the diagram above,
and more plain still from the annexed one, in which the instrument is seen in its position of elevation. It is of the greatest
consequence that the pulleys should be attached at the proper
joints, so that the long instrument may be in equilibrium, and
that the gestatory or bearing ropes may grip it with equal
power and hold it up uniformly at all points, for if the two
central pulleys b and c be attached too near to the rest F, the
tube will bend forwards to the ground at either end, showing
an upward convexity about the centre; whilst you find, on the
contrary, if you attach the two pulleys (b and c) farther off than
you should, that the ends will rise and show a convexity towards
the ground. So that it is a matter of nice skill when the instrument is being first adjusted that not only the pulleys (b and

c) but the others also (a and d) be so arranged and applied that
the bearing ropes may hold up the whole instrument with equal
power at every point, and support it in a straight line. And it
will be of great service to you in this respect, if you first
thoroughly ascertain on which side of the upright blocks,
whether in front or at the back, the pulleys are to be attached,
or which pulleys are to be attached on one side and which again
on the other. For unless you are thoroughly well informed on
all these points you will never bring the whole instrument to a
level, so that all the square perforated upright blocks (d) shall
be parallel with the plane of the horizon—but you will detect
many curvatures in consequence of its bending, although it is
not of much moment if the instrument has some slight turns
and bends in it, provided that it preserves a perfectly straight
line throughout its length. Yet it is best that the instrument
should not only have its horizontal line quite level, but also at
every point be true to the perpendicular. For you will so more
readily find any object than if the instrument were suspended in
an oblique or bent position. Accordingly I advise that you
take every pains on that first occasion to adjust the instrument
properly and most exactly, both as regards its horizontal level
and its altitude, which it will certainly be in your power to do,
if you only do not fear the trouble. How and at what points
all the pulleys as illustrated according to scale were attached to
my instrument, you will more readily learn from the engraving
itself than from a tedious explanation. First I fastened four
pulleys to the upright planks, i e., the first at a the foremost pair
of cramps, the second at b, the third at c, and the fourth at d,
over the wheels of which a very stout rope, called by me else-
where a gestatory rope, is passed in such a way that I passed
one end of the rope, first of all over d, then c, again over the
lower wheel of the double pulley e, then over the pulleys b and
a, and lastly a second time over the wheel of the double pulley,
this time the upper wheel, fastening the ends at f. The length
of the rope must be well adapted to the work, since if it is too
short, it by no means holds up the instrument with sufficient
power; again if it is longer than it should be so as to be slack

beyond the rest k, the tube is with very great difficulty main-
tained in its straight position, as the rest F would easily slip
from under the double pulley c and turn over to the great dis-
arrangement of the instrument; whereas if you keep this rope at
the proper length, the rest F can lean against this rope, or which
is the same thing, the rope passing twice through the double
pulley e, holds up the rest, so that it cannot turn over and quit
the mast, but the better preserves its position. So when this
double pulley e has been hoisted up by means of the guiding
rope y x, the gestatory rope d, c, e, b, a, e, f, feels and raises
up the instrument with perfect uniformity at all the proper
points, for the aforesaid rope passes with the greatest freedom
over all the wheels, and does not hitch anywhere. Nay, it most
readily yields to the double pulley e to suit different elevations,
as a person familiar with mechanics will readily understand.
Whereas if I had attached two different ropes, *i.e.* the one at b
and c, the other again at a and d, and had carried them only
over the two wheels e, the instrument could not have been
raised so equably, and settled at an exact level. For at one
time this, at another time that rope would have become now
slacker, now tauter, and this would have induced, immediately,
a slanting of the instrument. For I would have you know that
this instrument, of such vast size, on account of its excessive
length, though its parts be never so strongly fastened together
and constructed, yet very easily curves and quits the straight
line. Hence unless you take precautions against such a
disaster by hoisting up the instrument with perfect equability at
all points by means of ropes yielding to one another, passing
over different pulleys, you will find all your trouble a complete
farce and failure.

But be it remembered, though this particular rope passing
over the six wheels be perhaps the most important of all, yet it
is by no means sufficient of itself for the purpose of keeping the
instrument in the requisite straight line. For though it should
fulfil its office about the centre of the instrument at a, b, c, d,
yet it can never keep the ends g and i, from very apparently
bending towards the horizon. To prevent this therefore I

determined first to place the rest F standing on its base G about
the centre of the instrument, and to make it fast there by two
cramps, and to keep it steady by stays that it might not easily
turn over ; but how all these arrangements were completed and
put together I cannot very clearly demonstrate by a plan, or in
writing. You must therefore investigate what I have omitted
for yourself.

Then over the wheel of the support—the upper wheel k,
exactly as was done in the case of the former sixty-foot tube, I
passed another stoutish rope (which I may call the directory),
towards the end g of the instrument, and at the same point
where the weight requires it, or to a hook fixed there, I fastened
the end of the former rope ; the other after it had been fastened
round the lower extremity of the instrument, I firmly attached
to a winch at i, which was formed of a toothed lever about three
feet long, and a cogged axle, together with some other
mechanism. You must take care that the directory rope, if
new, is thoroughly stretched before using, so as not to swell or
expand too much afterwards. Since this rope g, y, b, is attached for the purpose of guiding both the upper and lower ends
of the instrument, and generally, to raise it sufficiently so that
they may recover any divergence from the straight line, and this
is very easy of execution. For by means of a handle on the iron
lever, which acts as an endless screw (to make it clear I have
given an illustration in the diagram above), the rope g y b is so
strained and drawn towards the winch, that both ends of the
instrument are forced to rise and become straight. And this is
effected with very little effort by one hand, by means of the
winch, so that if there is any error you can restore the whole
instrument to perfect straightness in an almost miraculous
manner ; and this too at any time, even after the instrument has
been hoisted and pointed upwards, whenever you please, if you
see that there is any need of correction and rectification. For
the winch is situated at the lower end of the tube, where it will
always be in your power to get at it, and turn the handle. In
this way and by means of the winch, after I have passed the
single rope from g through k, to b, it is in my power to check and

control this enormous and very weighty instrument one hundred
and forty, nay, one hundred and fifty feet long, so far as the ends
are concerned, so that if this instrument were fifty feet longer,
and extended to two hundred feet, such an instrument as I
trust I could equally well construct by the help of Providence ;
the guiding rope *g k b* would not be alone sufficient, but a much
longer rope would be required, which I should in like manner
have to pass over pulleys to be placed at the extremities of
the instrument, as we did with the intermediate primary rope,
and also over the lower wheel of the rest *F* ; and between the
two ends of the rope I should have to attach the winch or endless screw—the best appliance known up to the present time
to relieve the strain arising from the length and weight of the
telescope. Moreover, were this found insufficient to stand the
strain, you might pass an intermediate bearing rope, already
passing over four pulleys, over six, and also over the third wheel
of the support *F*, by all of which effective support would be
given to such an immense instrument. Now you can see in
short that there is no lack of means by which I could build and
guide such a monster. My present instrument, however, one
hundred and fifty feet long, requires no other means, nor any
longer ropes, since it can be excellently well steadied
and supported by those mentioned. And this I have
demonstrated. Briefly then so far as was possible, has
this account been given of the laying out and elevation of this telescope as regards the management of its
length. But you will say, "All is not cleared up by these
instructions, nor is your journey sped ;" for now that this
structure has been framed of parts formed by only two smoothly
planed boards ; admitting that no part is easily liable to lateral
curvature, yet nevertheless, when the four parts have been combined and united by the aid of the cramps, the instrument, from
its slender build, will beyond doubt very easily curve and bend
to some significant degree. I grant that the excessive length
and very slender build of the body of the instrument will lead to
some lateral curvature—for instance, when it is being hoisted
up, and is in process of being moved, especially if you direct the

whole instrument from one end; and this must, I allow, be
avoided with the utmost care if you are anxious to do any
accurate work with the tube.   "But how," you will say, "can
you prevent such a mishap?"   "Most easily," I reply, "and
that too by very light and simple apparatus," for though the
instrument be very weak where the extremities of its different
portions have been joined and bound together by the cramps
and wedges, I attach a well-stretched cord, strongly fastened,
at *m* as well as at *o*, passing it along from *p* to *r*, and then again
from *s* to *u*, so that the joints of the upright planks and the
cramps which bind them shall lie over the centre of the cord, at
which point I apply a deal rack, four feet long, but very slight,
an inch and a half broad, and one inch thick, to the smooth
plank forming the base of the instrument, as will be clear
from the preceding diagram (*A A*, *n g t*). Having fixed these
wooden racks to serve as checks on either side, the outside and
the inside (exactly as the cord was affixed at the same point in
the manner mentioned above), then at whatever point the
instrument shows a bend I pull and strain the cord to the
requisite degree, and by laying it on the handiest cog of the
racks, prevent it flying back under the strain (just as though we
were pulling a bow), and this must be done at every point
where there is any appearance of bending until you can clearly
see that the instrument has been brought back to a perfectly
straight line.   This is easily ascertained, even without those
sight adjustments which generally serve my tubes as *pinnacidia*, which I have mentioned in a preceding chapter.   If you
merely direct your eye along the instrument, by the tops of the
square diaphragms, any curvature will be immediately detected,
and should you notice that any error has been committed by
overstretching the rope—for example, at *n*—on one side, pull
directly the cord *v v* in the opposite direction, and correct the
excess of the last correction; and so do with the remaining
curvatures, one after the other, till you can obtain a perfectly clear
view through the whole line of diaphragms from one end to the
other of the instrument.   By this process you will not only restore
the instrument to a perfect level, but even strengthen it too as

regards the parts adjacent to the joints, and so much so that if you will take hold of the end where the eye-piece is placed, you will find no difficulty in moving at your pleasure the other, or object end, though the distance is so great ; the instrument all the while remaining straight and symmetrical; but it is necessary that you raise and handle the instrument whilst the weather is calm, for otherwise the wind is apt to do very great mischief to it, and keep it constantly rocking. You see, then, kind reader, how sometimes by a trifling thing of scarce any importance a great and difficult enterprise may be assisted, when proper means are rightly applied and at a seasonable time. You have an example here, where by bending a slender cord with one finger over a notched spoke, quite slender and fragile, we are able to manage and direct an immense mass by a trifling power, so that by inserting the cord in one or other of the projecting cogs you immediately perceive a significant change affecting the whole instrument ; and so the same being rectified in every part, and also a thorough examination having been made as to whether it has in any way lost its symmetry or is twisted, I fasten on the two boxes H and M, made of very light and thin boards, the former (H) to which the many-jointed tube is fastened by two connections, the latter (M) in which the object-glass is inserted. I insert these at each extremity of the instrument by wooden braces and pins, so that they can by no means even shake, much less fall. You must previously, however, well ascertain at what point, with respect to the distance between the lenses, M, with its convex lens, should be attached ; since, when the tube has once been hoisted, it would be a great labour then for the first time to investigate the distance, and so often to lower and again raise up an instrument of such size. Therefore it is better for you to make sure of and try the true interval by terrestrial objects whilst the machine is on the ground. When all these further processes and observations have been made the whole instrument is perfectly ready for hoisting and revealing phenomena. Seeing, however, that its bulk is so great that it extends far in one continuous length, and requires plenty of room where it may be put

together, set in order, and hoisted up, I cannot manage it in the
observatory attached to my house, chiefly by reason of the thickness and weight of the mast, which has to be erected piece by
piece.   I was forced, therefore, to choose another place and
receptacle on a small suburban estate not far from the town
where I dwelt, a magnificent piece of level ground, whence there
is a perfectly clear view all round.   A special inducement, too,
was that there are excellent facilities there for keeping under
cover both the instrument itself as well as the apparatus belonging to it.   On this spot, under a perfectly open sky, I erected
and built up a mast nearly ninety feet long, with a foundation
deeply set in the ground, and on a cross base.   It was built of
very stout planks, and had four cables to hold it in position, and
built strongly enough to bid defiance to and remain steady
against the violence of any storm.   In this mast a series of
horizontal holes was bored, to hold steps, so that the carpenter
or any other artisan employed might be able more conveniently
to ascend and fasten the pulley above mentioned, with its two
wheels, to the top of the mast, and afterwards to pass the long
thick guiding rope over the four wheels of the two pulleys x and
y, in order that the huge instrument might be moved as easily and
readily as possible, for the more numerous the wheels over which
the guiding rope passes the less the effort required to move and
raise weights, though they are thus moved a little more slowly.
When these arrangements have been made, it is then necessary
that the pulley y should be attached to the other double pulley
(e), over which the bearing rope passes, but this must be so done
that the rest k may remain with its own guiding rope (g, k, b)
between the pulleys mentioned (y and e) and the mast; for the
simple reason that the rest F may not lean to either side, and
so cause the entire instrument to deviate in the least from its
upright position; and this will be still further prevented if the
greater number of the pulleys (a, b, c, d) be attached to the side
of the mast, fronting us as we stand by the winch; yet this
must be done with very great care and judgment.   So then the
rest having its own guiding rope passed over and drawn taut on
the fourfold bearing rope, it can in no wise, along with the

instrument, be turned over, but is securely supported.   Moreover, I placed at the foot of the mast a wooden cylinder on
two upright posts.   The cylinder turns on pivots, and has two
levers through it crossing each other.   By means of these the
whole immense instrument is quickly hoisted up, as far as is
requisite for the observations, by only two men.   After it has
been elevated, and the small tube, made in two joints, and
usually fitted with two eye-pieces, has been screwed on to the
box (*II*), and well secured, according to my practice, with
clamps, I attach to the instrument that same movable telescopic table, the description of which has been given in the
preceding chapter, and fix it between two supports, as I mentioned in Chapter XX.   Then I attach to it two pulleys
furnished with a guiding rope and their proper balance weight,
not neglecting some other preliminaries.   It will now be in
your power to manage the largest tube which can be made—
this is one hundred and fifty feet long—with equal regularity,
readiness, and quickness, and to direct it to the smallest object,
just as if it were a tube of twenty feet or less, and with no
greater exertion, since the balance weight and the endless screw
have the same effect on the largest instrument as on a smaller
one ; for the entire machine is supported on its own centre of
gravity, except that the lower part has a designed preponde-
rance over the upper, in order that the balance or weight of
five or six pounds, which yet varies in different tubes, may keep
the instrument in equilibrium at any height.   Now, on the
subject of this telescopic table of mine and its use, since I have
spoken at great length already in the preceding chapter, I refrain
from saying any more here, but refer the reader back.   You
will perhaps exclaim, " You have indeed described to me an
instrument with two boxes attached, one at each end, for the
purpose of holding lenses, but in what way, I ask, is the entire
instrument and its intermediate parts to be closed up?   Will
the instrument you have described perform its duty, being quite
open on all sides?"   I reply that the instrument is certainly
not a perfectly closed tube, yet nevertheless it does the duty of
one quite as well as if it were really covered in on every side ; for

since the square perforated boards are not more than three or
four feet apart, they totally exclude light from the lenses and
from the eye of the observer, so that when you apply the eye to
H, which is the first hole, you will see nothing but the tube in
complete obscurity, and at the end a perfectly round hole; for
all the perforated tubes are completely blackened on the side
turned towards the observer, so that you catch nothing but the
tube with its veil of blackness. Had it been requisite, it would
have been easy for me to make a really round tube of this
instrument by inserting some light paper tubes three feet long
from hole to hole and from end to end, as in this present instrument of mine, beginning from the square box (*H*) as far as the
fourth hole. This you can see at *z z z;* and so you would
have a tube actually covered in on all sides. Besides, I could
have produced the same effect in another way, by covering the
entire instrument with very fine light black linen or silk. But
however easy it may be to do, we have no need of all this so long
as the first three or four spaces between the boards beginning at
the box *H* to about the length of ten feet be closed by small
tubes. As I said just now, we can very well do without all the
rest, as you will find to be the fact if you try. Lastly, I have
this piece of information to give about the holes, that as the
instrument is of very great length, the holes are made pur-
posely of unequal size; those in front a little smaller, those
behind progressively larger and larger, so that when you look
through they may appear perfectly equal, otherwise, had they
been quite the same size, those farthest off from the eye would
have appeared far the smaller; and therefore I gave the front
ones a diameter of eight inches only; those behind of nine, ten,
or eleven, although it would have signified very little had they
been really equal, provided the straightness of the instrument was
preserved so as to afford a perfectly uninterrupted view throughout.
Still, it is better that the first holes should be somewhat larger
than the succeeding ones, for fear the view should be inter-
fered with during some temporary bend in the instrument.
Besides, if the back holes are larger, it does not

much signify if there should be some slight flexure
of the tube about the centre. Lastly, as it is a work of great
labour, and one that needs no little expenditure both of money
and time, on account of the number of hands required to put
together this enormous instrument, to fasten it tightly, to fit it
with tackle, to hoist it up, to move and direct it; so also it is
no common labour to lower the same, to take off its fastenings,
to disjoint it, to loose the tackling, to put all the parts in their
right places, and keep them there; yet in truth it cannot be
otherwise in this matter, nor ought any astronomer to deem it
an irksome and troublesome business. For though the mast
may retain its place in the open field, the instrument with the
pulleys, ropes, and all the tackling cannot remain in the same
place, but it is quite necessary that all, and each, should be unfastened, removed, and laid up orderly in their appointed places;
otherwise all these articles obtained and put together at no
trifling expense and labour will be worn out and ruined by the
effects of bad weather. Now, in what way this calamity may
be avoided I hinted in the preceding chapter, and it was, if some
suitable place could be found and fitted up for holding the instruments and taking observations. But this is a matter which
not every private gentleman has it in his power to accomplish.
There should be some prince, some great Mæcenas of celestial
studies, most liberal with money, and the promoter of the whole
scheme. Under such conditions, I feel sure that an observatory
might be founded and fitted up to which access to the best constructed and most perfectly fitted telescopes shall be possible at
all seasons and in all weathers, as often as may be agreeable.
Where, whether the telescopes be twenty, forty, sixty, a
hundred, nay, a hundred and fifty feet long, you shall have
nothing more to do, as the guiding ropes will be always hanging
down ready for use, than to elevate any telescope you please,
nay, two or three of the largest size at once, and point them to
the stars; and again, whenever you please, at once to restore
them with all their tackling, whole and uninjured, to their proper
receptacle; a place where they may severally remain secure from

disturbance, and never be exposed to inclemency of climate; where they shall be so arranged as neither to interfere with one another, nor with the observations. I say nothing here of the many other advantages of such a place, but will treat of them all fully in the next chapter, and will show, by a clear diagram, that the scheme would be a success.

CHAPTER XXII.

Of a Special Observatory Suitable for the Largest Telescopes.

I FEEL fully persuaded that an observatory of the kind perfectly adapted for directing to the heavens and adjusting telescopes of great length can be well enough erected, and fitted with the necessary apparatus, by a method to be presently explained. But, as I have already hinted above, the enterprise is not one for a private person to undertake, but for some great nobleman possessing ample room, money, and, above all, enthusiasm, to promote so high an enterprise in the interests of astronomy. For, believe me, the work would be at first elaborate and expensive; but were it once in trim the chief drawbacks to the present work of observing would be thereupon obviated, so that we should always come to our work with every preparation complete, and should never have to rig up those long and weighty telescopes with so much loss of time and so much trouble.

Let us imagine, then, a raised tower on some very open and clear spot, not cone-shaped, but of equal thickness from top to bottom all round, built of beams rivetted together, or of bricks, for, as you will see, the structure must be piecework. The height should be one hundred or one hundred and twenty feet, if it is to serve for telescopes of one hundred and fifty feet long. Otherwise the height of the tower may certainly be less. The diameter may be at the architect's discretion, to suit the work to be carried on, and proportioned to it, although it should be as small as possible if no speciality has to be taken into consideration. For if regard is to be paid to the telescope only, it is sufficient if the diameter be no more than twelve or fifteen feet; as it would then not only be large enough to contain all the arrangements which, in my opinion, ought to be there, but also to admit of the construction of the different chambers and storeys. Now, that you may more fully and clearly understand my idea, I have done my best to present it distinctly to you in the subjoined illustration *B b*, from which it will be easy for any one possessing but a slight knowledge of mechanics, at once to thoroughly understand the whole matter. I would then build on a plain, or some very open spot of ground, a tower, such as you see at *A*; and I would erect all around it a quadrangular hoarding rising to a considerable height, resting on fixed wooden supports, so as to present the appearance of a procestrium (or open gallery) or of a very large stage. The sides of this, in order that they may reach to the length of the telescopes, should be at least one hundred and fifty feet long. With respect to the height of this stage, say fifteen or eighteen feet at most, so that the rests, which receive the centres of the telescopes and over which the building rope passes, may have room to stand. To form this, stout beams should be laid on posts and supports in sufficient numbers to bear the weight of the many frames of the tubes, and so that the whole stage may be covered in. In the flooring close to the tower an aperture should be made that there may be room for the telescopes stored up with all their tackling in the basement to pass freely when hauled out, and which may again be closed to prevent rain getting in, and for fear that any of the observers or spectators walking about in the dusk should fall in. You may fashion in the tower as many stages as you please and require. The basement should be the store-room of the telescopic apparatus. The rooms above may be kept for different purposes, as will be more fully stated presently. It is of the first necessity that the tower should be so constructed that the entire top, along with the four beams which support it, to which besides pulleys have to be attached, should be able to revolve, because the telescopes, though they may have been once fixed for observations, cannot be kept in one and the same position, nor at one and the same elevation, but must be turned and directed according to the variations in the motion and position of the object. Hence it is necessary that the covering of the tower should be so arranged that it may turn in any direction as easily as possible, and be also made to revolve without much difficulty by the aid of one or two assistants. In order, however, that you may understand all this more clearly, I have done my best to lay the whole matter before you by means of different diagrams. First, then, you must know that on the top flooring of the tower, where the roof springs, I place a very stout wooden wheel, peculiarly grooved inside, so as to have a ledge all round at the bottom, as you may see at No. 1. To the under-part four beams crossing at right angles are fixed, partly that the wooden wheel may be more firm, partly that the wooden cog-wheel (a) may be connected with the periphery of the wheel by means of a plank, b. At c c I hollow out the circumference of the wheel in order that another wooden wheel with cogs may be inserted, and turn round in the groove, c, c, as is shown in No. 2. And in order that this wheel may be rendered stronger, as it has to support not only the four cross beams (f f f f) from which the pulleys and ropes, as well as the great weight of the telescopes, depend, but the entire top also, I further strengthen the wheel, d, with four cross pieces, e, e, on which the rafters of a roof generally rest, so that the entire top, along with the supporting beams, telescopes, and all the apparatus, can be turned round by the aid of one or two men. Now, in order that this may be more easily managed, I fit the small cog-wheel, a, in such a position that it may properly bite the cogs of d, and so turn the wheel this way or that. The cog-wheel (a) does not require any winch in the upper storey (H or G) to turn it, but it is immaterial where the power is applied. Here, in the annexed figure (B b), the wheel (a) is well worked from the chamber (C), where I place for this purpose a cylinder, turning on pivots, fixed in a frame of cross beams, supported by uprights. To this cylinder or windlass, which turns on its own supports, another little cog-wheel (b) is attached, to be worked by the large crown wheel (g), through the centre of which the end of the tall mast (F) which revolves in its bracket (k) passes, like the point of a spindle. The other end of the upright mast passing upwards through all the storeys supports the upper cog-wheel (a), so that the large wheel before mentioned (d), together with the supporting beams (f f), as well as the transverse beams (e e e e), on which the supports to the roof rest may be moved and turned round. This is accomplished by a handle or bar (m) without much trouble, so that we can at any moment, with the greatest

ease and quickness, direct and turn to every quarter of the
heavens even the largest and heaviest telescopes; and if you
prefer to have a shorter mast for this purpose it is better to
transfer the winch from the chamber (C) to one of the upper
storeys (G or H). To raise from the ground and support in the
air the telescopes themselves I place a lifting engine, specially
constructed of four blocks of wood, by which the telescopes can
be simultaneously lifted up, as you may see in the storey E, and
still more clearly in plate No. 4. For as pulleys with their ropes
hang from four different beams it is necessary that there should
be four lifting engines to raise and direct each tube; but the
guiding rope is passed over the pulley (f) again (n), so that the
whole work can be performed under cover, with the exception of
the observations, which are made by the observer in person, in
the open air, on the great stage. And now, as already mentioned, in the spacious covered room, telescopes of any size,
together with all their apparatus and all their tackling, can be
conveniently kept safe ; and, again, when they are wanted, as
often as you please, whatever be their length, by merely lowering
the bearing ropes from the top beams and attaching to them the
telescopes, can be raised from the vault or covered chamber, and
placed in position for observation, according to requirements,
without much trouble. You can then direct to the stars not
one only, but four telescopes all together; nay, more, if need
were, provided you attach to the wheel (d) a corresponding
number of bearing ropes. As to the means by which each telescope is to be drawn into position by ropes, as well as managed
according to your wish by the aid of a guiding table, I need
make no recapitulation here. I have enlarged upon all these
points in their proper places, so that nothing more remains but
to inform you that when the observations are finished all you
have to do is to pass the telescopes through the apertures or
doors (B and I) into the vault, and to take off the ropes in the
upper storey, and finally to cover the aperture with its lid, and
then you have stowed away and placed in safety all and each in
such a manner that they are not only safe from rust and injury
by weather, but you have them all in complete order whenever

you have to return to your observations. Thus we can avoid the wearisome trouble of being obliged first of all to grope out these enormous instruments from all sorts of cellars, to put them together, to fasten and unite them, first this piece, then that, to make them tight and keep them so, then rig them up with such a number of ropes, pulleys, and winches; and then shortly afterwards, when the observations are finished, which have lasted but a few hours, to unfasten, dismantle, and carry them away, to place them in their receptacles, with very great trouble to the observers, and, what is most important, with great loss of time. For which reasons, that is on account of the great trouble and expense of these operations, many excellent observations are abandoned and omitted by observers which would otherwise never perish or pass away unmade, if a perfectly convenient observatory were available, to the great exaltation and benefit of astronomy. But it may be objected, "Where in the world would you find a man endowed with sufficient means to construct such a building?" for as I am forced to own to myself, even the tower takes a Crœsus to build it. That you may, however, see that it is not altogether impossible for men of moderate fortune as me and you, I will now unfold a still simpler plan which we can carry out with far less expense to ourselves, scarcely less convenient and useful than the grand and costly tower I have described with its spacious open stage or procestrium. Here is the plan:—You must construct at the foot of a mast, such as is described in Chapter XXI., on some open spot wherever you should think convenient, a chamber covered in all round by means of posts and boards, of such length and breadth that telescopes about a hundred and fifty feet long, if you fancy that length, may be stored in it and kept covered; but the chamber must be entirely closed in, and the top well boarded over, except that there should be a certain number of apertures in the flooring, in order that in the stage above (B and I) there may be a perfectly free passage for the telescopes. A chamber such as this would in my opinion suffice for storing the telescopes and all the necessary apparatus, so that you would scarcely have anything to desire on this score except that it would be necessary on the occasion of every observation to loose the guiding rope, with its pulleys, from the upright mast.   It cannot be denied, therefore that by this method we avoid very many difficulties and troubles; but I must add that an observation so taken becomes yet easier and more convenient, if you dig a trench near the mast of a size to contain a crypt or covered gallery of sufficient length, breadth, and depth for the storing and safe keeping of the telescopes.   You must protect the sides of the crypt with boards, or brick them.   The upper part should be closely covered with boards to the level of the ground, but there must be apertures left, through which the telescopes can be drawn out and again put back.   In this way you would have all you could desire or expect even from the costly tower above pictured, for you would be able to stow away all the apparatus in the subterranean covered gallery made at far less expense, so that there could be nothing more to do than to uncover the apertures, to attach the ropes to the mast, and then to feast your eyes on the sight of tubes of various kinds all prepared and equipped.   Meanwhile, kind reader, if the plan of the observatory which I have disclosed does not happen to please your fancy, if you will, as a worker in the service of our goddess Urania, discover some better, and disclose it to me, you will merit the warmest thanks of all astronomers.

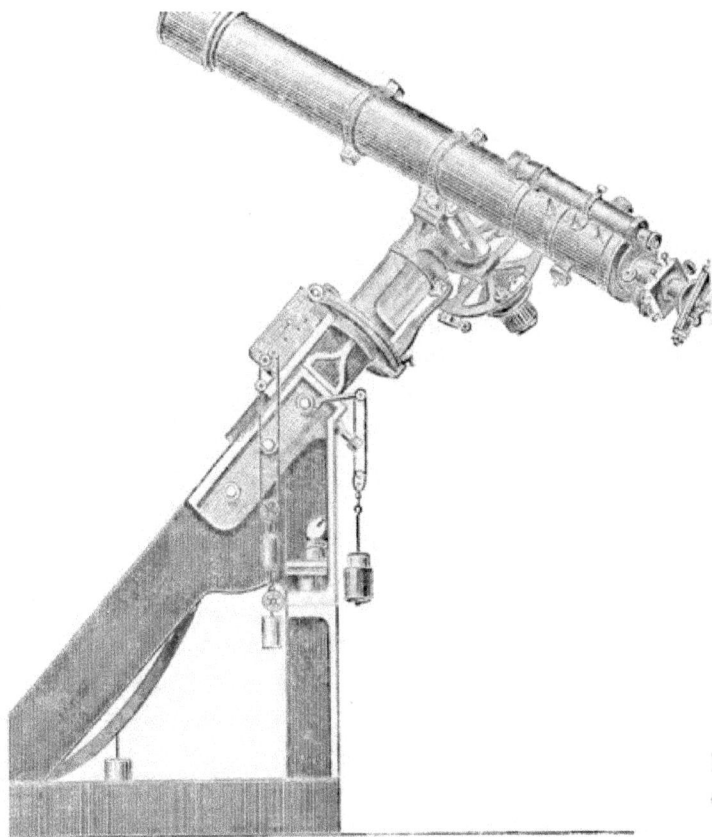

"Hoc sub pace vacat tantum: juvat ire per altum
Aëra, et immenso spatiantem vivere cœlo,
Signaque et adversos stellarum noscere cursus."

THE TULLEY EQUATOREAL
AT CROWBOROUGH OBSERVATORY.

NOTES.

I WILL now offer some remarks upon the several engravings.

The portrait of Hevelius (see frontispiece) was printed, so far as I can ascertain, when he was in the sixty-ninth year of his age. I have copied it from the second volume of his " Machina Celestis."

Plate 1 represents his original design for sketching the various phenomena connected with the sun's surface. This contrivance was superseded by that shown at

Plate 2, fig. W, cap. xviii., which represents Hevelius, assisted by his friend Ismael Bullialdus, making a sketch of a solar eclipse. He must have spent many a day in arranging and completing this very complicated machinery for overcoming the difficulty of following the path of the sun's motion in right ascension, and declination, during the continuance of any phenomenon. The idea of elevating the axis of his instrument to an angle equal to the complement of the latitude, upon the meridian, does not appear to have occurred to him. How many of his difficulties would have vanished, could he have employed an equatorial mounting of the present day, may be recognised by an inspection of that of my principal telescope (see Plate 7 on preceding page), and which to every practical astronomer, who has been accustomed to its use, renders a nonequatorial mounting simply intolerable. Nevertheless, its excellence and convenience consists mainly in its simplicity, as may be gathered from the description of it which I shall append.

Plate 3. This plate shows some preliminary arrangements for the erection of his great telescope, all of which may be easily understood by reference to his detailed account. Some of the citizens are taking an interest in the operations, and watching the adventurous carpenter ascending the mast. It gives a view of part of the city, and the men employed at their several duties. The houses appear of very uniform construction, built chiefly of wood, and roofed with earthen tiles.

Plate 4, fig. Aa, represents the arrangements for actual obser-

vation almost completed, and the general interest in the scene
increases. Some magnates have arrived, and others are arriving,
from various quarters in their carriages. An important personal
introduction is proceeding; perhaps King Ludovicus XIV.
has arrived, to whom Hevelius dedicates his book. The
astronomer is standing at the eye-end of his telescope and
watching, with anxious interest, the final adjustments of his
laborious work. It is very difficult to imagine how Hevelius
could suppose that such an apparatus, for carrying his lenses,
could be erected free from great flexure and consequent distor-
tion of any object under examination. That the early telescopes must have been sadly wanting in rigidity, and
general steadiness, is amply shown by reference to the
two following diagrams of planets as seen and depicted by
Hevelius and Gassendi respectively. Diagram A represents

DIAGRAM A.

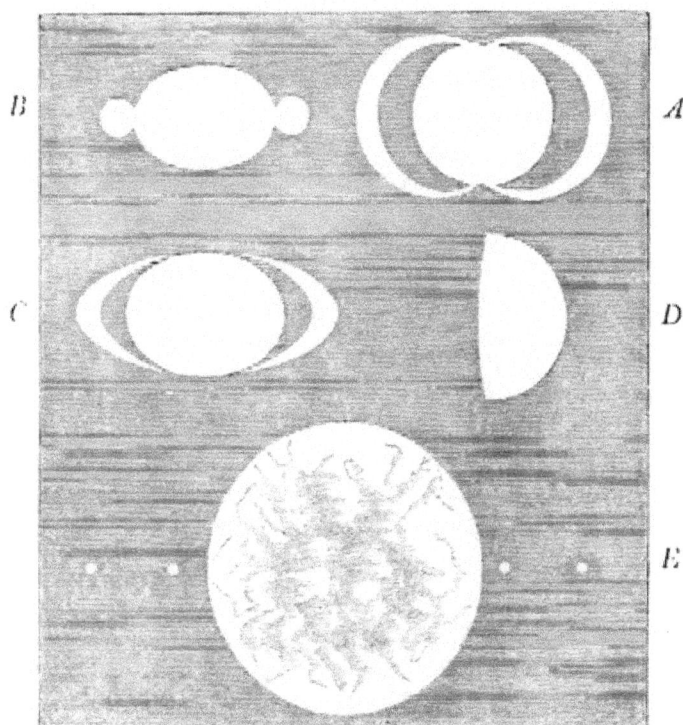

drawings of Saturn (*A, B, C*), Mars (*D*), and Jupiter (*E*), by
Hevelius, which I have copied from his "Selenographia," for,
curiously enough, he does not give any representations of them
in his larger work.

Inter alia—he thus speaks of Saturn after having carefully observed the planet—

"Quod igitur primum ad Saturnum attinet, is jam inde usque
â multâ antiquitate ipso oculorum testimonio non refragante,
rotundus habitus est; postquam autem oculo armato inspectari
cœpit, ovalis ferè, sicut Kepplerus et alii mathematici contes-
tantur, apparuit. At simul atque hoc instrumentum opticum
perfectius est redditum, de die in diem, per Telescopium majis
elaboratum inspectus, non solum ovalis, sed et in utroque latere
duobus adhærentibus globulis præditus apparuit, ita ut ex tribus
partibus compositus videatur, quemadmodum ex præsente figura *B*
manifestum est. Hâc specie Saturnum sæpenumero conspexi
utpote anno 1643 mense Octobri et Novembri ; duo adhærec entes
globuli utplurimum Eclipticæ parallelæ erant. Totus autem sum
in hâc opinione quod non semper Saturnus oblongus et quidem
duobus parvis globulis acuminatus appareat, sed quod interdum
hi globuli post Saturnum latitent, quasi duæ stellæ (et quidem
definitis temporibus) quæ Saturnam circumeunt. Memini
namque, quod ipsum, mense Septembri et Octobri anno 1642
plane rotundum conspexerim et quidem distinctis vicibus.
Eandem figuram in Saturno quoque observavit Summus Philoso-
phus et Astronomus P. Gassendus eodem anno, mense
angusto, ut extat in ejus judicio de novem stellis circa Jovem
visis pag. 14. Atque refert ibidem, quod Galilæus ante 30
annos, cum quoque tali figurâ rotundâ suo telescopio specta-
verit. Adhæc de eâdem stellâ Saturni commemorat Matthias
Hirschgarter, in suâ detectione dioptricâ quòd Primarius et
Nobillissimus Vir, cui nomen Fontana (sicut ex aliis illud nomen
didici) Neapolis, egregio quodam tubo hunc planetam inspexerit,
eumque planè aliâ et diversa, quam cæteros, formâ observaverit ;
siquidem in quolibet latere, loco supradictorum globulorum,
ansulam bisectam adjacentem vidit, ita ut quælibet à vero corpore
distincta, sensu oculi armati percepta sit, et per cujuslibet ansulæ

cavitatem prospectus in cæruleum cælum ac æthera patuerit;
veluti apud præfatum Autorem pag. 22 scriptum legitur, Ejusmodi faciem Saturni in præsens diagramma *A* consignavi.
Hæc relatio de stellâ Saturni multis perquam admirabilis et vix
credibilis videtur; nec ego diffiteor, quòd initio me ab assensu
sustinuerim; postquam autem longiores ac meliores tubos mihi
comparavi, et per eos Saturnum inspexi, hanc jam descriptam
faciem Saturni non merum somnium esse, sed magnam partem
ita cerni reapse deprehendi. In nonnullis tamen partibus ali-
quantum diversam ejus faciem animadverti ab eâ, quam paulò
ante designavi. Medium enim, idemque maximum corpus
Saturni, in oblongiori formâ mihi apparuit; Brachiola quoque
utriusque lateris, ex parte alia mihi visa sunt : siquidem illa inter
se, cum medio corpore Saturni adeò aretè non cohærebant, sed
ubi in unum continuum Corpus coire et cohærescere debebant,
in tam acutam et exilem cuspidem definebant, ut non percipi
posset, quod non oblongo Saturni corpore strictè copularentur;
præterea spatium, quo Brachiola ab àpso Saturni corpore aliquo
modo separabantur, per quæ cæruleum cælum licebat intueri,
non æquabat istam latitudinem, quàm prior figura repræsentat,
sed minus erat. Insuper, quod in exprimendâ verâ hujus
planetæ formâ maximè dignum est animadversione, uterque
arcus, tam interior, quam exterior, brachia terminans, nequaquam
sectionem circuli, ut ab Eximio Fontana annotatum; sed para-
bolicam, seu potiùs hyperbolicam sectionem refert; sicut ex
figurâ *C* cognoscitur. Hanc enim veram esse Saturni faciem,
longo et exquisitæ operæ tubo accuratè intueri, omniaque probe
considerare potui, ita ut unusquisque, qui cupiditate reperiendi
veri ducitur; huic indefessæ observationi tutò possit fidere."
    With respect to the diagram of Jupiter, he says (inter alia)
" In stellâ Jovis etiam occurrunt non unius generis memora-
bilia, ope telescopiorum præstantiorum observata. Globus
equidem Jovis non insuetâ et peregrinâ formâ, instar Saturni,
sed satis rotundus conspicitur; nihilo tamen minùs deprehen-
dere licet, cum non esse omnibus numeris orbiculatum, neque
politum instar tornati ac lævigati globi; quoniam si illum per

PETRUS GASSENDUS, DINIENSIS.

———

Hic est Ille, dedit cui se Natura videndam,
Et Sophia æternas cui reseravit opes :
Invida non totum rapuistis Sidera Vultum
Nantolius, Mentem pagina docta refert.

tubos meos (quorum adminiculo diameter ejus sex, imó veró septem feré digitos æquat) inspecto, discum ipsius minus radiosum, atque magnis certisque maculis conspersum, ad exemplum feré Lunæ reperio, quæ cæteris partibus longè obscuriores cernuntur.

"Hanc figuram Jovis perquam diligenter, quoad fieri potuit, in diagrammate antecedente, penes *E* volui exprimere. Propter immensam autem distantiam Jovis à Terrâ hæ maculæ nondum telescopiis hactenus usitatis, etiamsi præstantissima sint, internosci et à se invicem (prout Maculæ Lunæ) distingui potuerunt. Intereà speramus telescopia longè perfectiora, ex sectionibus Conicis hyperbolicis in medium allatum iri, quæ figuram formamque cælestium corporum adhuc apertitis, illustriusque detectura sint."

With respect to Mars, he says (inter alia)—

"Adhæc planè mihi persuadeo hanc planetam, ceu corpus aliquod opacum, sui luminis admittere vicissitudines, instar Veneris, Mercurii et Lunæ, ita tamen, ut nunquam possit conspici corniculatus vel falcatus, more reliquiorum inferiorum; sed phasin bisectam obtinere, quando est perigæus et in quadrato Solis versatur, sicut Kepplerus idem statuit in epitome Astronomiæ Copernicanæ pag. 843. Etinim quod hæc sententia non solùm sit probabilis, sed et ipsi consentiat experientiæ, optimo telescopio deprehendi, anno 1645 die 26 martii horâ septimâ vespertinâ, sicut et die 28 ejusdem, ubi maximam partem dimidiatus apparebat, sicut phasin ejus delineavi in præcedente figura *D*."

Diagram *B* represents some drawings of Saturn as seen by Gassendi between the years 1633 and 1656, and the following are his remarks respecting them, which I have copied from Vol. IV. of his works. These are interesting as showing probably the first drawings of the planet after the discovery of the ring. I may here mention that I sent, some time since, a copy of these drawings to the Royal Astronomical Society, and that they were published in the Society's Monthly Notices for the year 1876, p. 108.

DIAGRAM B.

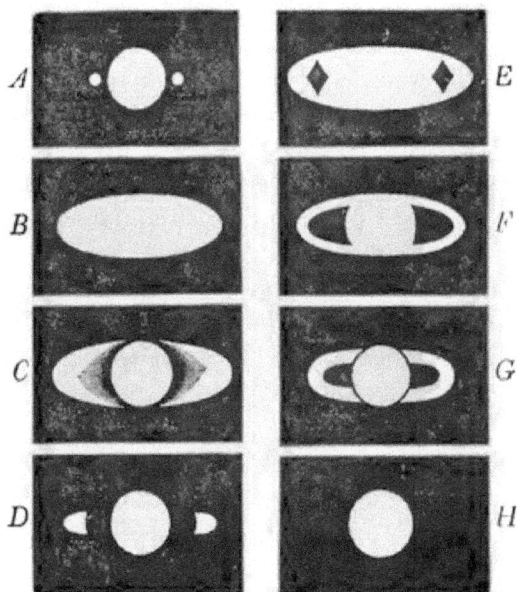

Figure *A.* June 19th, 1633.—" Postea circiter decimam cum per varios nubium hiatus Saturnum tubo respicerem, is quasi ovum sericum, seu quo bombyx filo deducto concluditur. Diameter longior (existens ferè secundum longitudinem Zodiaci) vix apparuit minor diametro Venerea, utraque nempe visa est repetita octies aut decies adæquatum proxime diametrum foraminis tubi. Et a parte quidem anteriore|ansa, seu appendicula ostensa est confusior; sed a posteriore exhibita est omnino distinctè; totumque hâc prope magnitudine et formâ conspectum est; siquidem interdum corpus Saturni rotundum, neque radiis undique ansas complectentibus visum est; interdum vero cum ipsis ansis ob circum effusos concinnos confusiùs."

Figure *B.* April 13th, 1634.—" Saturnus telescopio maximo visus est oblongus, et qualis semper aliis."

Figure *C.* Nov. 20th, 1636.—" Attendere placuit ad formam Saturni, eaque exhibita fuit, non quasi adjunctis orbi medio duobus aliis orbiculis, sed quasi adnatis duabus ansulis interceptis maculis, quasi foraminibus effictæ. Heinc forma ovallina, et ea sanè longiuscula adeò ut medius quasi nucleus vix superaret trientem totius longitudinis. Fuit autem nonnihil clarior, candicantiorque ipsis ansis. Longitudo semper protensa secundum eclipticam. Habes utcunque heic effigiatum. Diameter martis apparuit minor sensibiliter diametro brevior Saturni."

Figure D. Jan. 11th, 1645.—"Visi sunt adhuc distinctiùs Saturni satellites quasi duo cuculli hac propemodum specie Jovis verò cum Mediceis ita se habuit."

Figure E. March 18th, 1646.—"Cùm ad Saturnum telescopio attendissem medius ille circulus albus non est mihi visus planè exquisitus speciesque fuit propè hujusmodi."

Figure F. December 8th, 1650.—"Saturnus hujusmodi ferè fuit."

Figure G. Nov. 21st, 1651.—"Saturnus sic se propè habuit."

Figure H. Jan. 16th, 1656.—"Vesperi Parisiis Saturnus Rotundus apparuit sine satellitibus clarissimo viro Ismaeli Bullialdo, quemadmodum etiam solitarius deprehensus fuit ab ipsomet Amanuensi mensibus Februario ac Junio telescopio majore Dygbeano videlicet, et minore Galileano."

Having given, in the above diagrams, some engravings of the planet Saturn from the works of Hevelius and Gassendi, I purpose giving another representation of the planet, drawn by Hevelius, nearly thirty years after those (A B C) in diagram A; as well as one by Ball, Hooke, and Cassini, respectively, which represent the current ideas of the general form of Saturn between the years 1664 and 1676. I will first notice the drawing by Ball as being the earliest in order of date. It is a somewhat extraordinary circumstance that, during the last 40 or 50 years, Mr. Wm. Ball and his brother, Dr. Ball, have been credited with the honour of being the discoverers of the primary division in Saturn's ring, and which has been styled by many eminent astronomers up to the year 1882 as "Ball's Division."

In the year 1880 it occurred to my friend, W. T. Lynn, Esq., F.R.A.S., to refer to the original communication of the Messrs. Ball, to the Royal Society. In doing so, he found no positive conclusion could be drawn about their assumed discovery, without their drawing, which for some unaccountable reason was not given in the " Philosophical Transactions," although he tells me he was at the pains to consult three copies of Vol. I. Mr. Lynn did not prosecute the enquiry any further until he paid me a visit on Sept. 15th, 1882 ; when we discussed the subject, and I stated to him that for many years I had been impressed with the idea that Cassini—not Ball—had discovered the primary division in the ring of Saturn ; and I showed him Lowthorp's " Abridgement " of the early volumes of the " Philosophical Transactions " as my authority for the supposition. In this volume Mr. Lynn found Ball's drawing of the planet, which is wanting in the " Phil. Trans.," and of which I give a reproduction, as well as a *verbatim* copy of the original communication, which was as follows :—

"This observation was made by Mr. William Ball, accompanied by his brother, Dr. Ball, Oct. 13, 1665, at Mainhead, near Exeter, in Devonshire, with a very good telescope near 38 feet long, and a double eyeglass, as the observer himself takes notice, adding that he never saw that Planet more distinct. The observation is represented by figure 3, concerning which the Author saith in his letter to a friend as follows : This appeared to me, the present figure of Saturn, somewhat otherwise than I expected, thinking it would have been decreasing, but I found it full as ever, and a little hollow above and below. Whereupon the Person to whom notice was sent hereof, examining this shape, hath by letters desired the worthy Author of the System of this Planet, that he would now attentively consider the present figure of his anses or ring, to see whether the appearance be to him as in this figure, and consequently whether he there meets with nothing, that may make him think, that it is not *one* body of a *circular figure* that embraces his *disk* but *two*. And to the end that other curious men, in other places might be engaged to joyn their observations with him, to see whether they can find the like appearance to that represented here, especially such Notches and Hollowness as at A. B. it was thought fit to insert here this newly related Account."

It is not stated who the "Person" was to whom this communication and the drawing was sent; but in all probability it was either Wallis, Hooke, or Huyghens. The annexed copy of Ball's drawing of the planet I have copied from Lowthorp :—

A

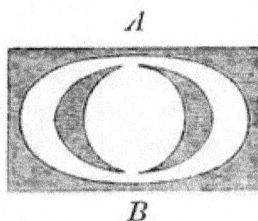

B

Now, in this plate, not the slightest trace of any division in the ring is to be seen, and I consider that a very great misapprehension has arisen as to the precise meaning of the following words : "Whether he there meets with nothing that may make him think that it is not *one* body of a circular figure that embraces his disc but *two.*" I am disposed to think that he meant the two ansæ; for what reference could "Notches and Hollowness" have to a *line* of division on the ring?

The next engraving, in order of date, is Hooke's, of which the following is a representation :—

It refers to an observation by him on June 29, 1666, viz. :— "Between 11 and 12 at night I observed the body of Saturn through a 60 foot telescope, and found it *exactly* of the shape represented in the figure. The ring appeared of a somewhat brighter light than the body; and the black lines *a a* crossing the ring, and *b b* crossing the body (whether shadows or not I dispute not) were plainly visible; whence I could manifestly see that the southermost part of the ring was on this side of the body, and the northern part behind or covered by the body."

In Hooke's drawing no trace of a division in the ring is perceptible, which would surely have been the case had he supposed that Ball had really made such a very important discovery. The lines *a a b b* which he describes as crossing the ring are fully accounted for in the explanation of the plate, and can only be intended to show his ideas as to the shadow of the planet falling upon the ring, and that of the ring upon the planet.

The next engraving is by Hevelius, nearly thirty years after he made the drawings *A B C* in diagram *A*.

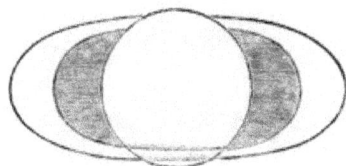

It is interesting, as showing how much better he had been enabled to inform himself of the relatively true form of Saturn and its ring. Had Ball really discovered the division Hevelius was not a man who would have omitted, either to satisfy himself of the truth of such discovery, or to have omitted its insertion into his own drawing upon its confirmation, and I must maintain that, up to the year 1675, the division in Saturn's ring had not been discovered.

The next and most important engraving relating to this enquiry is by Cassini, and bears date August, 1676.

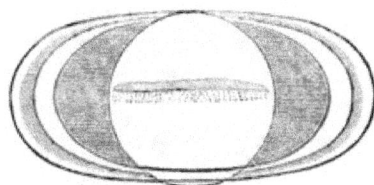

We have here, for the first time, an actual drawing of the division in the ring, and his communication to the Royal Society runs thus :—

"Ex Schemate Saturni à Hevelio ante annum observato video, cum Telescopiis nostris longè inferioribus, uti. Tunc enim temporis (ut et nunc Aug 1676) cernebatur nobis in Saturni Globo Zona subobscura, paulò Australior centro, instar Zonarum Jovialium. Deinde latitudo Annuli dividebatur bifariam, Lineâ obscurâ apparenter Ellipticâ revera Circulari quasi in duos annulos concentricos, quorum interior exteriori lucidior erat. Hanc phasim statim post emersionem Saturni è Solaribus radiis per annum usque ad ejus immersionem conspexi ; primo quidem, Telescopio Pedum 35, deinde minori, Pedum 20."

The above appears to me a plain and decided announcement of the discovery, and makes no allusion whatever to any previous observation of the kind.

A confirmation of Cassini's discovery may be found in Dr. Smith's Optics, 4to., 1738, Vol. II., p. 440, from which the following is an extract :—

"In the year 1676, after Saturn had emerged from the Sun's rays, Sig: Cassini saw him in the morning twilight with a darkish belt upon his globe, parallel to the long axis of his ring, as usual. But what was most remarkable the broadside of the ring was bisected quite round by a dark elliptical line, dividing it, as it were, into two rings, of which the inner ring appeared brighter than the outer one, with nearly the like difference in brightness, as between that of silver polished and unpolished ; which, *though never observed before*,* was seen many times after, with tubes of 34 and 20 feet ; and more evidently in the twilight, or moonlight, than in a darker sky."

I am informed that there are other astronomical works which attribute to Cassini the discovery of the division in Saturn's ring ; but during the present century, Kitchener in 1825, Smyth in 1844, Hind in 1852, Proctor in his "Saturn," p. 49, Breen in "Planetary Worlds," p. 217, &c., all repeat the story of the brothers Ball having discovered this feature in the ring of Saturn. I have made this inquiry in the hope of elucidating the truth, and of recovering, for Cassini, the honour which would appear to be due to him.

Plate 5, Figure B *b*, discloses his plan for the erection of an observatory, suitable for telescopes of any size, and he gives some minute details of its construction. It does not appear, however, from this, or any subsequent account, that the proposed building was ever erected under his superintendence ; nevertheless the author deserves some praise for the ingenuity displayed in

* The italics are mine.

the arrangements. I should consider that, had he erected it, he would not have met with the practical success which he had anticipated. It will be observed that the two smaller telescopes have closed tubes, the benefit of which he would doubtless soon appreciate. Had he arranged for the rotation of the whole tower, instead of merely the roof, he would have found it much more convenient though probably ineffectual. His great fault, in all the chief arrangements, was the confidence he placed in the bearings of the ropes, which would be ever fluctuating with every hygrometric condition of the atmosphere.

Plate 6. This engraving represents a quadrant and sextant with which Hevelius determined, for the most part, the positions of the various astronomical subjects under his consideration. With these instruments he did not employ any lenses whatever, but trusted solely to naked eye observations through small apertures placed on the limb and at the radial centre of his instrument, respectively. Although the astronomers of that day doubted the accuracy of this apparently primitive method, yet upon subsequent examination of these observations by competent persons, they were found to be exceedingly accurate and trustworthy. His persistence in employing this plan led at first to remonstrances from, and subsequently to angry correspondence with, some of his contemporaries.

By constant practice, however, Hevelius had acquired such wonderful skill and confidence in his eye and hand, that remonstrance and opposition to his method were alike unheeded by him.

He used not only quadrants and sextants, but also octants ; the latter being a form of astronomical appliance scarcely ever made since his day.

The ornamentation applied to these several instruments was very elaborate. Upon the construction of the sextant, shown in the engraving, he appears to have composed the two following verses :—

> Herculis hic virtus, humeri poscuntur Atlantis
> Cælica si penitùs visere tecta velis
> Non hic pervigiles noctes, æstumve, geluve,
> Curis cum variis, sit tollerare grave.
> Blanda sed Uranie, quae splendet sola laborem
> Æternæ laudis præmia grata levant

PLATE 6.

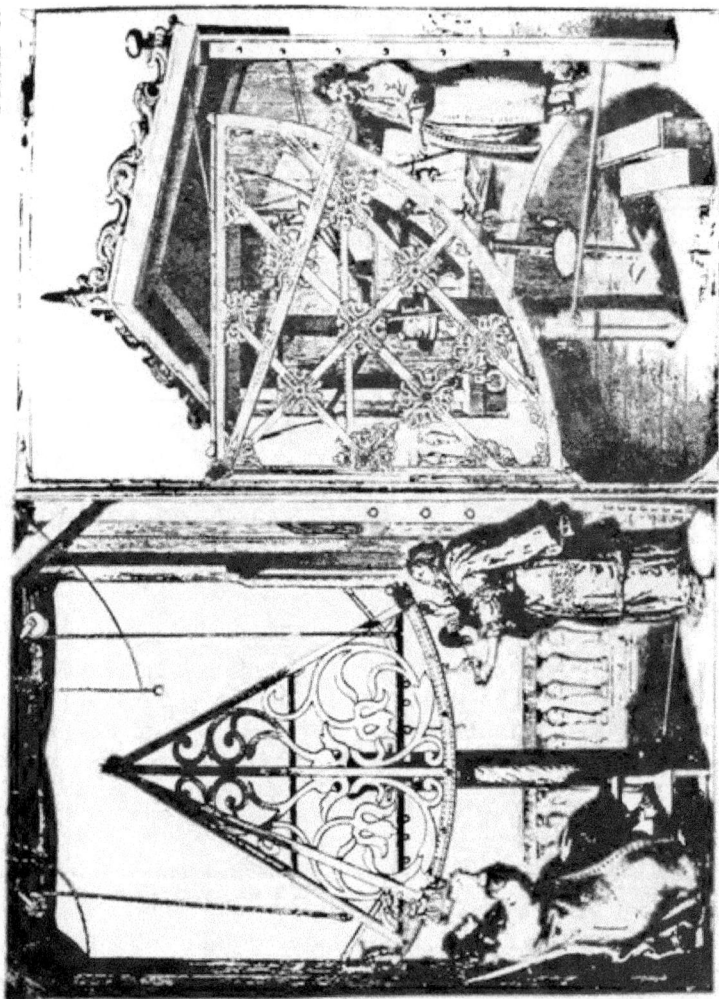

Ætherios varios subeunt quos fornice motus
Sydera, quosve solent cuncta tenere situs,
Promptiùs expandit solus quàm cætera *Sextans*
Organa, quæ solers fingere cura potest.
Ne molem metuas nimiam, pondusque molestum,
Artis subsidio quod levat una manus.

The following inscription was engraved upon it :—

Hunc Sextantem
Rei Astronomicæ bono, ex solido metallo,
Novâ et singulari planè ratione fabricandum
Curavit,
Johannes Hevelius
Cons. ac. p. t. JUD. V. C.
Gedani
Anno æræ Christianæ 1658.

Respecting a large octant he writes as follows :—

"Quò verò etiam post fata nostra, si hicce noster præstantis-simus octans in plurimos annos, ut benè meretur, integer fortè conservabitur, successores quoque nostri non nesciant, in cujus gloriam, quo fine, tempore, atque Auctore conditus sit, hanc inscriptionem loco instrumenti congruenti cælo æri incidere placuit.

In honorem
Supremi Architecti
ad augendas
Res Astronomicas
Nocturnis stellarum contemplationibus
Quod felix, et faustum sit!
Octantem hunc æneum
destinavit
Johannes Hevelius
Cons. V. C.
Gedani
Anno 1659."

THE TELESCOPES AT THE OBSERVATORY, CROWBOROUGH.

Latitude North .	. . .	51° 3′ 14″
Longitude East .	. . .	9′ 30″
Ditto ditto (in time) .	.	0 38″

HAVING concluded my remarks upon the foregoing plates, I will now give a description of the two equatorial telescopes at my own observatory.

Plate 7 exhibits my chief instrument, which has some historical associations connected with it. It was constructed originally for the Royal Astronomical Society of London, and formed the subject of a report of a Committee appointed by its Council for the purpose of examining its merits. Full particulars of this report may be seen in the " Memoirs of the Society," Vol. II., pt. 2, p. 507.

At that date (1826) it would appear that extreme difficulty was experienced in procuring flint glass of even small dimensions, which circumstance had a tendency to impede the progress of practical astronomy. At length a disc of $7\frac{1}{4}$ inches in diameter was placed at the disposal of the Society by Messrs. Guinand and Regnier, of Neufchatel, to be examined and reported upon according to its merits. A disc of this size was probably, at that time, unique in England, and the Council placed it in the hands of Messrs. Dollond and Tulley, with directions to take every proper means for ascertaining its efficiency for optical uses ; and it was finally agreed that Mr. Tulley should undertake to form it into the concave lens of an achromatic object-glass of twelve feet focal length. Mr. Tulley after considerable difficulty finished his task, and the performance of the completed telescope proved in the highest degree satisfactory. The report concludes as follows :—

" The light of this telescope is, however, amply sufficient for showing the nebulæ of Sir Wm. Herschel's first class. Several of these were examined, and the high degree of concentration of the rays in the focus, arising from the absence of aberration, proved very valuable, and was evidently marked in the resolvable appearance exhibited by them. Saturn was shown with great distinctness, the division in the ring (b), and the three interior of the five old satellites being plainly seen. A satellite on the body of Jupiter was also seen, as well as its shadow; and the planetary discs of the other satellites could not be mistaken for spurious ones.

" Your Committee consider the facts above detailed (c) speak sufficiently for themselves as to the excellence of the telescope to render comments or praise on their part superfluous ; but they cannot close this report without observing once more on the great pains bestowed on its workmanship by Mr. Tulley, and his address in availing himself of the resources of his art in operating on a material which might certainly in the beginning be regarded as highly unpromising.

" (Signed)   G. DOLLOND,
J. F. W. HERSCHEL,
WM. PEARSON."

The aperture of the object-glass is exactly 6·8 inches, and its focal length twelve feet.

From the Astronomical Society the instrument passed into the hands of the Rev. Wm. Pearson, LL.D., author of " Practical Astronomy," in which work are several micrometrical tables calculated for use with this instrument. At Dr. Pearson's death it was bought, at his sale, by the late A. Ross, of Holborn, and from him, shortly afterwards, by Captain Wm. Noble, F.R.A.S., of Forest Lodge, Maresfield, Sussex, of whom I purchased it in the year 1855. Dr. Pearson mounted it upon a parallactic ladder, which was ill adapted for an instrument of its size. Soon after it came into my possession I mounted it upon a peculiar form of stand, made in

b The inner (crape) ring and the division in the outer ring are very well shown under favourable atmospheric conditions.

c These details included the examination of some difficult double stars.

London, from drawings, and under the superintendence of
Frederick Brodie, Esq., F.R.A.S. It is of cast-iron, in three
pieces, each bolted firmly together, having lead between the
joints to lessen any liability to tremor. The bottom part is
formed as a triangular plate. The two bearings at the base of
the plate rest upon a large baulk of timber, twelve inches
square. Previously to this baulk being placed in position it was
sawn through its entire length, and one half reversed. An
iron flitch, twenty-one feet long, eight inches deep, and one
inch thick, was then placed between the two lengths of
timber, and the three fastened firmly together by ten screw bolts.
The wall, upon which this support rests, is two feet in thickness
and about forty feet from the ground. A piece of lead, rather
larger than the united baulk, was placed beneath it in order to
lessen vibration. At the base of the eastern end, of the
north side of the triangular plate, is a large iron nipple, which
bears upon a shallow depression in an iron plate, let into a stone,
which rests upon the main support reaching from wall to wall.

At the base, on the western end, is a large vertically placed
screw, the point of which also bears upon an iron plate let into
a stone, which rests upon the main support. This screw will
either elevate or depress the western angle of the stand, for the
purpose of levelling the declination axis.

The apex of the triangular plate rests upon a large iron jaw
placed upon the southern main support, and on either side of
this jaw, placed at right angles to it, are two massive screws,
by means of which the whole stand is shifted in azimuth.

The southern support is a plain timber baulk of the same
size as that at the base, and without an iron flitch, on account
of the bearing being much shorter between the walls. The floor
of the observatory is entirely independent, and does not touch
any part of the stand. The declination axis is $8\frac{1}{2}$ feet from the
floor. The northern side of this stand tapers from 3ft. 6in. at
the bottom, to a width of 6 inches at the top. The main adjusting
screws for latitude and azimuth are attached to the bottom part
of the frame. The polar axis is also bolted to the stand at the
upper part, having two taper bearings bushed with brass—dia-

meter of the bottom, 3¼ inches; top one, 2½ inches. A cast-iron cradle turns on this axis, the bearings of which are also bushed with brass. To this is attached the hour circle, while the top of this cradle carries the declination axis. On the top of the polar axis, and inside, at the top of the cradle, are two corresponding discs of steel, upon which the whole weight of the telescope cradle and declination axis rests, so that the cradle turns with a very small amount of friction. There is an adjusting screw which acts on the discs, so as to lift the cradle from the taper bearings just enough to allow it to turn freely without lateral motion. The declination axis is made of brass, and is hollow; the end covering the cradle of telescope has a bearing of 2½ inches diameter; the other end has a bearing of 2 inches diameter, and carries the declination circle and counterbalance. This circle is 18 inches diameter, and is divided on silver to ten minutes, reading to ten seconds, by verniers. The hour circle is 15 inches diameter, divided on silver to two minutes, reading to four seconds, by vernier. On the polar axis is fixed the clockwork and tangent screw. The weight of the stand is about three-quarters of a ton, and the weight of the moving parts, together with the tube, renders the telescope remarkably steady, and perfectly free from oscillation of any kind. One great advantage of this form of stands arises from the facility with which you can observe the pole-star, but with the temporary inconvenience of being obliged to shift the clockwork; as this is attached to the polar axis by two screws only the movement is easily effected. There are several minor adjusting screws in several parts of the stand which need not be particularised here.

Situated on the top of Crowborough Hill, which on clear nights is generally free from the vapours which form on the lower ground, the good defining qualities of the instrument are fully apparent. When I first commenced my observations here the improvement in the brightness of some rather faint stars was so marked that I began to suspect some of them to be variable. This increase in brightness, however, was entirely due to the clear atmosphere of the hill.

The instrument is supplied with a parallel line micrometer, a double image micrometer, a battery of eye pieces ranging between powers 15 and 800, and a star spectroscope.

Plate 8. Represents my portable equatorial telescope which was made about the year 1773, by Mr. Jesse Ramsden for Sir George Shuckburgh, Bart., and was given to me by his grand-daughter, the late Lady Cath. Vernon Harcourt, of Buxted Park, Sussex.

This very compact instrument is an excellent example of Ramsden's skill, and can be made available for any use to which such an instrument need be applied, while its graduation for right ascension, declination, and azimuth, was completed in the most perfect manner. In the " Philosophical Transactions " for the year 1793 Sir George Shuckburgh thus alludes to this telescope :—

" However, after some years had elapsed, the idea of an equatorial telescope was again renewed by three several artists in this kingdom, Messrs. Ramsden, Nairne, and Dollond, with many very material improvements, such as to carry the portable equatorial almost to perfection. Of this instrument Mr. Ramsden had made three or four as early, I believe, as the year 1770 or 1773; viz., one for the late Earl of Bute, one for Mr. McKensie, another for Sir Joseph Banks, and lastly one for myself, with which I made a great many astronomical and geometrical observations in France and Italy in the years 1774 and 1775, some of which may be seen in a Memoir upon the heights of some of the Alps printed in the " Phil. Trans." for 1777. Of this machine a plate and description in French was printed in the year 1773, and reprinted in English in 1779."

The principal parts of this instrument are :—

1. The azimuth circle representing the horizon of the place of observation, which moves on a vertical axis.

2. The hour-circle representing the equator placed at right angles *to*, and moving *upon*, the polar axis which represents the axis of the earth.

3. The semi-circle of declination (on which the telescope is

PLATE 8.

THE RAMSDEN EQUATOREAL AT
CROWBOROUGH OBSERVATORY,
ADJUSTED TO THE LATITUDE OF 51° 3′ NORTH.

placed) moves on what is called the axis of declination, or the axis of motion of the line of collimation.

4. The telescope is an achromatic refractor, with a triple object glass, the focal distance whereof is 15 inches, and its aperture 2½ inches. There are six different eye pieces. By a contrivance in this equatorial the telescope may be brought parallel to the polar axis, so as to point to the pole star in whatever part of its apparent diurnal revolution it be.

5. The refraction apparatus (for correcting the error in altitude occasioned by refraction) goes on upon the eye-end of the telescope, and consists of the following parts :—

1. A slide which moves in a groove, and carries the eye tubes of the telescope; this slide has an index upon it corresponding to five small divisions engraved on the grooved plate.

2. A very small circle, called the refraction circle, movable by a finger screw at the extremity of the eye piece. This circle is divided to half-minutes, but only numbered to minutes ; one entire revolution of this circle is equal to 3' 18". The moving of this refraction circle raises the centre of the cross wires on a circle of altitude.

3. A quadrant of 1½ inches with graduation for altitude. To this quadrant is attached a small round level, which is adjusted partly by pinion and partly by the index of the quadrant.

The principal uses for which this equatorial is applicable are—

1st. To find your meridian by one observation only.

To do this, elevate the equatorial (or hour) circle to the co-latitude of the place, and set the declination semi-circle to the sun's declination for the day, and hour of the day, required ; then move the azimuth and hour-circles, both at the same time, either in the same direction or the contrary, till you bring the centre of the cross wires in the telescope exactly to cover the centre of the sun ; that being done, the index of the hour-circle will give you the apparent or solar time at the instant of observation ; thus you get the time, though the sun be at a distance from the meridian, then turn the hour-circle till the index points precisely at twelve o'clock, and lower the telescope to the horizon in order to observe some point *there* in the centre of your eye-piece, and that point is your meridian mark found by *one observation only*. The best time of the day for this operation of finding your meridian, is three hours before, or three hours after, 12 at noon.

2nd. To point the telescope on a star, though not on the meridian, in the day-time.

Having elevated the hour-circle to the co-latitude of the place and set the declination semi-circle to the star's declination, move the index of the hour-circle till it shall point to the precise time that the star is then distant from the meridian, and the star will be visible in the telescope.

These two examples will show that this form of equatorial is adapted for all the purposes to which the principal astronomical instruments (viz., a transit, a quadrant, and an equal altitude instrument) are applied.

This instrument is also supplied with a finder, which is fixed upon the azimuthal plate.

My transit instrument was made by Mr. Edward Troughton, sen., for Sir George Shuckburg, Bart. It has an aperture of 1¾ inches, and a focal length of 20 inches. I have also at the Observatory, for terrestrial purposes, a 3 inch achromatic telescope, by Wray, a 6 inch Transit Theodolite, by Negretti and Zambra, and several minor instruments.

C. L. PRINCE.

October 6th, 1882.

PRINTED AT THE "SUSSEX ADVERTISER" OFFICE, LEWES.

www.ingramcontent.com/pod-product-compliance
Lightning Source LLC
Chambersburg PA
CBHW021944190326
41519CB00009B/1142